THE CIVILIZATION STACK

THE CIVILIZATION STACK

Power, Orbit, and Compute
in Musk's New Industrial Machine

REID J. OSTERLEY

THE CIVILIZATION STACK
Power, Orbit, and Compute in Musk's New Industrial Machine

Published by Supercritical Books
New York, United States
www.supercriticalbooks.com

Library of Congress Control Number: 2026943511
Paperback ISBN: 978-1-968988-65-4
Hardcover ISBN: 978-1-968988-66-1
eISBN: 978-1-968988-67-8

Printed in the United States of America
First Edition 2026

For permissions or media inquiries, contact:
editorial@supercriticalbooks.com

Contents

PART I

THE ORBITAL LAYER

1

Friday at Starbase

THE PAD BEFORE LAUNCH

ON MAY 22, 2026, STARSHIP Version 3 stood on Pad 2 at Starbase, on the strip of Texas coast where dune grass, mudflat, tank farm and launch towers had been crowded into a few hard-used acres. At 5:30 p.m. Central Time, it would lift off. The vehicle rose in unpainted stainless, longer through the tanks and more severe in profile than the ships before it, its metal skin exposed because paint rides upward too and every needless pound leaves the pad as a tax on the mission. The vehicle kept changing from flight to flight, the tanks lengthened for propellant, the engines below revised for harder work, the plumbing pulled tighter inside the hull, each iteration bringing the flying machine closer to the cargo ship the program meant to send past orbit. Seams, frost bands, access ports and the engine skirt made the machine visible as a sequence of built and tested pieces.

This was Pad 2. For most of the program's life a single launch mount at Boca Chica had taken every Starship, one vehicle at a time, each flight held up by the damage the last one had left behind. A launch does real violence to the thing it stands on. Thirty-three engines lighting a few feet above the deck drove heat and pressure and debris straight back down, and in the early years the first full-stack liftoff tore into the launch mount, blasting concrete and debris outward and leaving a crater beneath it. With one pad, the whole site moved at the speed of that repair, a rocket spending a few seconds trying to destroy the table it launched from and a crew spending the next days putting it back. That bottleneck was what Pad 2 was built to break. With its own tower and its own deluge, it gave the program a second line, two vehicles working through stacking and test and flight without waiting on each other, and this flight was the first to leave the new pad.

Crews filled the site that day. A great many had never held a job anywhere else. They had come out of universities and out of other people's failed rocket companies and, increasingly, out of high schools in the Rio Grande Valley. They had learned the trade on this stretch of sand. No aerospace town had stood here before SpaceX built one. The company had made the place and then made the people for it, training them on its own hardware by crews it had trained the same way a few years earlier. The shift schedules ran around the clock, and the housing the company had thrown up nearby had turned a fishing hamlet into something closer to a company town, a few hundred people whose working lives were organized entirely around the cadence of the pad. North and west of the pads the site had filled in over the years with the buildings that cadence required, the high bay where the stages were stacked, the long production sheds where the steel rings were rolled and welded, the two launch towers standing over the flats with their catch arms folded. Across the flats the whole complex read as a working factory that happened to point at the sky. The tank farm north of the mount held spheres of liquid oxygen and liquid methane, their steel walls sweating frost. The propellant ran cold enough that loading it into the vehicle would shrink the steel and pull the whole stack in on itself by inches before it ever moved. The loading had to hit full tanks and full chill within a window of minutes before the clock. Vapor curled off the vents and slid downhill toward the water. Out past the cordon, onlookers gathered in the usual turnouts along the highway, and the live feed carried the flight far beyond the Texas coast.

For the people running the flight, the vehicle first arrived as numbers on screens. Engine temperatures fell toward the band where turbopumps could accept ignition. Tank pressures climbed their ramps; valves changed state. McGregor had forced Raptor from design into behavior one firing at a time, until an engine was known by its limits under heat. At launch, the teams watched those limits in pressures, temperatures and currents, each station responsible for one part of the vehicle. The flight computer would run the final sequence, while people watched for a condition the software had not been told to expect. Ten years before, this had been empty sand. The maps still named the nearest town Brownsville. The sign at the gate now read Starbase, and Starbase had become an incorporated Texas city the previous spring.

IGNITION

All thirty-three Raptor 3 engines on the Super Heavy booster ignited. The sequence had been rehearsed through testing and earlier flights, but this was the first time the new vehicle and its engines had flown together. The ignition had to march out across the cluster in a fraction of a second, each engine confirming itself lit and stable before the count would commit the vehicle to leaving the ground, and any engine that failed to come up cleanly in that window could still scrub the flight with the booster already roaring on the mount. Beneath the deck the water deluge fired at full pressure, tens of thousands of gallons a second thrown up into the path of the exhaust to swallow the acoustic energy before it could bounce off the concrete and shake the vehicle to pieces. Hold-down clamps kept the vehicle pinned to the mount through the first part of full thrust, long enough for the flight computer to confirm that every engine it needed was lit and holding, and only then did the clamps let it go. For an instant the entire base of the stack vanished into a white cloud of flash-boiled water and burning methane. Then the rocket came up out of it and climbed over the Gulf.

The launch was also a measurement run. Range cameras followed the ascent, while instruments in the engines, tanks and flight computers recorded how the vehicle behaved under load. The measurements could be read against the design, and against the next vehicle already taking shape at the site. Two modified Starlink satellites rode aboard the upper stage for a different job, to image its return from outside the ship. The rocket carried instruments for judging the rocket that would follow it.

At a viewing site down the beach, the light reached spectators before the sound did. They saw the ignition with the vehicle already climbing on top of it, and for several seconds the shore stayed quiet while the sound crossed the distance between them. Then it came across the flats. It arrived as pressure before it arrived as noise, a physical wall that pushed against the sternum, rattled the temporary stands, set the car alarms going in the distant lot, and rolled through the crowd as the low frequency of thirty-three engines reached the body a beat before the ear could organize it into anything. The crackle came layered over it, the sharp tearing sound that big rockets make, loud enough at this distance to register in the chest as much as in the ears. People who had watched dozens of these still flinched when it hit. The crowd was larger than it had any right to be for a test flight that might end in the water, families and off-shift engineers and the standing population of people who had reorganized their weekends around watching rockets fly. Starship V3 rose over the Gulf while the sound lagged behind it.

Once clear of the tower, the climb became a sequence the regulars could read. The vehicle rolled onto its heading and tipped east over the water, setting its long axis along the path it would ride to space. It pushed up through the dense low air toward the point of greatest stress, the squeeze the engineers called max-q, where the speed of the vehicle and the thickness of the air combined into the hardest load the airframe would take. The engines throttled down for a stretch to ease the strain, then came back to full power once the air thinned out behind it. The plume changed shape as it climbed, the exhaust spreading wider in the thinning atmosphere until it stood off the back of the vehicle in a great translucent fan.

THE ENGINES THAT FAILED

During ascent, one of the booster's thirty-three Raptors shut down. The other engines kept the vehicle on its planned course, and the upper stage separated and continued its burn. One of that stage's vacuum engines also shut down, yet it reached the trajectory assigned to it. The two engine losses were on different vehicles, each with its own task still to finish.

A cluster of thirty-three engines gives a booster room to absorb an engine loss and continue climbing. Falcon 9 had flown for years on nine engines with the same kind of tolerance built into its ascent. Many engines made repeatedly on a production line gave a vehicle a way to survive a failure that would end a flight built around only a few large ones. Flight 12 put that capacity to work in the booster, while the upper stage continued after losing a vacuum engine of its own. The record from each stage could tell engineers how the vehicle held its course, where the margins remained and what had to change before another flight.

THE MACHINE BEHIND THE ROCKET

The vehicle that left Pad 2 carried one company's name on every document the launch required, yet it belonged to a much larger industrial machine. Start with the engines. The thirty-three Raptors were designed and built by SpaceX. At McGregor, in central Texas, the company puts Raptor through firing tests on dedicated stands; the rolling thunder of those tests has worked its way into the ordinary weather of the county. Raptor uses full-flow staged combustion, the configuration propulsion textbooks long treated as theoretically ideal and practically punishing, a cycle that sends every bit of propellant through the turbomachinery and runs that machinery hot, at pressures that destroy engines not designed to the inch. The Soviet program chased the idea for decades and never flew it. That cycle, at Raptor's scale, made the engine both a propulsion problem and a production problem. The latest version had been driven toward manufacturing as hard as performance, with part count cut, welds reduced, and labor hours forced down, because an engine you mean to build by the thousand is a different industrial object than an engine you mean to build by the dozen. A run of engines cheap enough to build in volume is also a run cheap enough to lose one from in flight, and Flight 12 showed the value of that redundancy when a booster engine shut down during ascent and the vehicle continued.

Take the satellites the rocket carried. Two modified Starlinks flew with the upper stage to record its reentry from outside the ship. They came from a constellation of thousands SpaceX had built and launched on its own rockets, a manufacturing line that could make more than communications relays. A rocket could put an instrument beside the next ship on its return and bring the image home to the people building the heat shield after it.

Go nearly four hundred miles inland, to Grimes County, where SpaceX and Tesla had named a proposed compute-manufacturing project with Intel's technical help. To make circuits there would take clean rooms, tools that lay down and cut patterns, process engineers who could hold those steps within tolerance, and a dependable supply of power and water. For now, the county name marked the next kind of work the builders hoped to bring closer to the machines they made.

The geography of the working places told the same story as the launch. McGregor supplied the engines, Hawthorne had spent years building rockets and capsules at the plant on the edge of Los Angeles where the company had started, and here at Boca Chica the Starships were welded and stacked within sight of the pad. Starlink satellites came off a line of their own. Each site had people who could make, test and change its part of the system, then send the result into another crew's work.

At the site, the Raptors were what hung beneath the booster and the satellites what flew alongside the returning ship. The working pieces mattered because one could serve another: an engine brought a satellite to orbit, and a satellite could return a picture of the heat shield the next ship would need. A launch was beginning to leave behind tools for the launch after it.

THE DIVIDED RESULT

The booster did not come home. After separation it pitched around for the boostback burn that should have set it on a course toward recovery, but not all the engines planned for that burn ignited. The vehicle struck the Gulf hard and was lost. Five days later, on May 27, the Federal Aviation Administration designated the flight a mishap and opened the review required before the next Starship flight. The work of tracing the boostback failure belonged to engineers and to a regulator charged with keeping the consequences of a test flight away from people on the ground. Flying hardware until it broke produced measurements; flying it again required an account of what broke and why.

The upper stage reached its planned suborbital trajectory despite losing one vacuum engine during its burn. It released twenty Starlink mass simulators and two modified Starlink satellites, all on the suborbital path. The simulators exercised the deployment mechanism that later flights would need for real batches of satellites. The two imaging satellites flew to watch the stage's reentry from outside the ship. Thousands of ceramic tiles had to stay attached to a flexing steel hull through that return. The stage completed its reentry and the planned splashdown in the Indian Ocean.

The flight produced a divided result. The booster failed its recovery and was lost to the Gulf; the upper stage reached its trajectory, deployed its payloads and returned through the atmosphere. The imaging satellites added a way to observe that return from outside the vehicle, while the failed boostback left pressure, thrust and guidance records for the people who would change the booster. The hardware in the water was gone. Measurements from both stages could move back up the line toward the next vehicle, standing half-built in the high bay while this one was still on the floor of the Gulf.

THE LINEAGE

A booster falling into the sea would have been the end of most rocket programs. It had very nearly been the end of this one, more than once, in the years when the company was small and the failures came with no margin behind them to absorb the cost.

The lineage was long, it was public, and most of it survives on video. The first rocket the company ever built, the small Falcon 1, failed three times before it reached orbit. The first flight was lost to a fire in its first stage about half a minute into the climb. The second lost control when its propellant sloshed in the tanks at the wrong moment and the engine shut down early. The third watched its spent first stage push up on residual thrust into the second after they parted and wreck the flight, and after that third loss the company had money for one more attempt. The fourth flight reached orbit. It was the first privately built liquid-fueled rocket to do so, and it arrived with the company a single failure away from the end. Years later came CRS-7, a Falcon 9 carrying supplies toward the space station that broke apart in the Florida sky when a single strut gave way inside the upper stage. The year after that brought AMOS-6, a rocket and its satellite both lost on the pad during a routine fueling test, an explosion that threw debris across the complex and took a paying customer's spacecraft into the fireball with it. There was the long, stubborn campaign to land a first stage on a drone ship in the Atlantic, attempt after attempt that ended with the booster coming down hard and folding itself into a fireball on the deck, their explosions caught on video and watched around the world, until one booster came down and stood there on the deck and did not fall over. There was the row of Starship prototypes that detonated through 2020 and into 2021, the squat test articles that flew a few kilometers up and came back down too fast and went up in flame, one after another, until at last one set down and stayed whole. Then came the first full Starship that tore itself apart over the Gulf a few minutes off the pad, and the later one whose launch tower reached out two enormous arms and caught the descending booster clean out of the air.

The company kept losing hardware, right through the Flight 12 booster. A loss could still stop the program while crews and regulators worked out what had happened. The useful difference was that a flight left measurements and a partly built successor. Engineers could compare the failed part with the trace, change a valve, a control rule or a weld, and test whether the change held on the next vehicle. The next launch had to earn its place from what the last one taught.

That method ultimately bought them cadence. By the middle of the 2020s the company was flying its Falcon rockets on a schedule no other organization on Earth came within reach of. In 2025 it flew 165 Falcon 9 missions, more than any other launch operator. It flew from Boca Chica and from two pads at Cape Canaveral and from Vandenberg on the California coast, sometimes within hours of one another, a rocket lifting off somewhere on a timetable closer to a busy regional airport's than to anything the space business had ever run. The drone-ship landings that had once ended in flame had become so ordinary that the broadcasts barely paused on them, the boosters flown and reflown until the flying was the cheap part of the whole enterprise, single airframes racking up flight counts that would have read as science fiction a decade earlier. What had looked, flight by flight, like a string of costly accidents had compounded into the highest-volume launch operation in the history of the world, and it had arrived there by feeding each accident straight back into the vehicle that flew next.

Other vehicles stood at other pads, with crews and schedules of their own. The loss entered the test record while the larger launch operation kept its capacity to fly. It moved on to the next vehicle, the way it had moved on from earlier losses.

Stand far enough back from one launch and the question grows beyond whether its booster came home. The engines can be made again, a pad can take another vehicle, and the satellites already overhead can watch a ship return. People who build each piece now have another flight's measurements to work from. How did access to space become repeatable enough that these places could begin to work on one another's problems?

2

Breaking the Launch Ceiling

THE DRONE SHIP AT 2AM

AT TWO IN THE MORNING, SIX hundred kilometers off the Atlantic coast of Florida, a Falcon 9 first stage stood on the steel deck of the drone ship *Of Course I Still Love You* as a tug began the slow tow toward Port Canaveral. It had landed there three hours earlier, dropping out of the night on a narrow flame, shedding the last of its speed in the landing burn, and setting its legs inside the circle painted with the ship's name. During the landing the drone ship held its own place in the swell on azimuth thrusters, an empty steel platform keeping station for a returning booster. The nearest people were the recovery crew on a support vessel a few miles away, outside the safety zone, waiting for clearance to close back in once the deck was quiet. The stage had come down at sea because of the job it had just done. A booster that throws a heavy payload toward orbit spends its speed climbing away from the coast, out over the Atlantic, with too little propellant left to reverse course and fly back to land, so SpaceX sent the landing pad to the place where the rocket still had enough margin to meet it. By the time the support ship came alongside and the crew boarded, the engines were cold, the booster scorched black down its windward side where reentry had marked it, the four titanium grid fins at its top stained bronze and violet by heat, the landing legs splayed and locked against the deck. The crew moved around its base with flashlights and tie downs, making the stage part of the ship for the long tow back. At seven to ten knots, the drone ship would spend roughly a day and a half behind the tug, carrying the booster back as freight from a place no runway could have reached.

Back at the port, a crane would lift it off the deck and lay it horizontal on a transporter, and it would go into a hangar where a crew would inspect it, swap what needed swapping, and put it back in the flight rotation. The whole value of the scorch marks was that they were survivable. A booster that came home could be looked over and flown again, and a booster that could be flown again counted on the company's books as an asset with flights left in it, where every rocket before it had been a cost the mission absorbed and discarded into the sea.

Their work was the whole proposition made physical. They washed the salt off, pulled the access panels, and put borescopes down into the nine engines to look for cracks the fire might have started. They replaced the parts that a flight was known to consume, the seals and the soft components, and left in place the great expensive bones of the vehicle, the tanks and the engine structures, that a flight did not. Then they stood the booster on a test stand and fired its engines for a few seconds to confirm it would behave, and if it did, the stage went back into the rotation with a flight or two of wear logged against it and many more still to give.

That calm was earned the expensive way. The grid fins that steered the booster down through the upper atmosphere had started out as aluminum, and the aluminum had caught fire and warped in the heat of reentry, so the company recast them in single pieces of titanium that could take the fire without ablating away, each one large enough to lie down on. The drone ship itself had been a graveyard before it was a parking lot. Earlier boosters had come in too fast and toppled, or set down hard and buckled a leg and tipped, or touched the deck and then burst across it in a roll of orange flame that the company posted publicly without much comment. Each wreck taught the next attempt something specific, a guidance law or a throttle profile or a leg that needed to lock harder, and the lesson was fed straight back into the booster that flew next. By the time a stage could come home to a barge at two in the morning with no one watching, the coming home had been paid for many times over in the wrecks that came before it.

For most of the space age, a launch vehicle climbed once, spent its stages into the ocean or burned them up in the atmosphere, and the next mission required a new vehicle built from raw metal. The Shuttle was a costly exception: its orbiter returned, but it took an immense workforce to ready it again. Now this booster had stopped being remarkable to the people at the port. There was no crowd waiting at the dock. There was a refurbishment crew on the morning shift, a hangar with the stage's name already on the schedule, and a flight assignment a few weeks out. Recovery had passed into the day's work order.

WHAT IT USED TO COST

Putting a single kilogram of anything into low Earth orbit aboard the Space Shuttle had cost something on the order of fifty-four thousand dollars. A kilogram of water. A kilogram of bolts. A kilogram of a human being, multiplied out across the mass of the astronaut and the seat and the suit and the air to breathe. The Shuttle had been sold to the world as the rocket that would make space cheap by flying again and again, and it did fly again and again, and it stayed ruinously expensive throughout, because bringing one back and making it safe to fly took an army of technicians months of work that ate most of what reuse was supposed to save. The orbiter came home sheathed in tens of thousands of ceramic tiles, every one shaped for its own place on the hull, and after each flight technicians went over the vehicle by hand, pulling and checking and regluing the tiles that the heat of reentry had cracked or torn loose. The three main engines were marvels of their generation, and they were unbolted and torn down to their components between missions. The two tall white boosters that flanked the stack at liftoff were parachuted into the Atlantic and recovered by ship, and salting a segmented steel motor in seawater and rebuilding it back to flight cost about what casting a fresh one would have. A standing workforce of thousands had to be kept ready to process each vehicle whether the program flew six times in a year or twice. The Shuttle had promised the turnaround of an airliner and delivered the turnaround of a warship in for overhaul. When it retired, the cost of access to orbit redistributed across a field of expendable rockets that had always assumed the Shuttle's price was the price of doing the impossible. An Atlas V brought a kilogram down toward eight thousand dollars. A Russian Proton, flown hard and sold aggressively into the commercial market, came in lower again, nearer three or four thousand. These were the good options.

Almost none of the launch price was fuel. The propellant that filled a rocket cost a rounding error against the price of a launch, liquid oxygen and kerosene against a vehicle that ran into the tens of millions. The expense was the machine, the engines and the tanks and the avionics and the labor of assembling them, all of it built to fly exactly once and then fall away. An airline that scrapped a jet after a single flight from New York to London and welded a new one for the return would charge fares no traveler could pay, and those fares would have almost nothing to do with the price of jet fuel. The space business had run on that model since its first day, and the people inside it had stopped seeing it as a model at all.

The price of orbit had been set in part by a particular way of building and discarding rockets. A crew that recovered a stage could inspect the expensive machine itself instead of ordering its replacement in raw metal.

The raw materials in an orbital rocket, the aerospace aluminum and carbon fiber, the liquid oxygen and the kerosene, the metals in the engines and the avionics, came to something like two percent of what a finished launch vehicle cost. The other ninety-eight percent was labor and tooling and testing and qualification and overhead, and above all the assumption that the vehicle would fly once and be thrown away. That ratio of finished cost to raw materials, the idiot index, runs absurdly high on a rocket, far higher than on a car or an airplane. A price built that way is a property of the production system, and change the system and the price moves with it. Reusability and a high flight cadence were the two levers built to move it.

The assumption held for reasons that went beyond habit. For most of the industry's history the rockets that mattered were built under contracts that paid a company its costs plus a guaranteed margin on top, and a business paid a fixed percentage over its costs has little reason to drive those costs down, since every dollar it saves is a dollar of its own fee surrendered. The customers were governments buying a handful of launches a year, each carrying a payload worth far more than the rocket under it, and for a customer like that the reliability of the launch mattered more than its price, and there was no volume of flights across which a cheaper rocket could ever pay back the cost of designing one. The whole structure, the contracts and the customers and the cadence together, rewarded getting it right over getting it cheap. It had selected, across decades, for companies that were superb at the first and had simply never been asked for the second.

Tory Bruno, who ran United Launch Alliance, pursued a different calculation as the landings began. Inspection, teardown and replacement could consume the savings from a returned stage. A flown stage also had to prove its reliability before anyone entrusted it with a crew or a costly national-security satellite. ULA's proposed answer was to recover the engines of its next rocket while letting the rest go. The different schemes asked the same practical question: which parts could be brought back, checked, and flown again often enough that recovery paid for itself? The answer would come from the flight logs and the work in the hangar.

THE FIRST LANDING

The booster that started it came down on the night of December 21, 2015, at Cape Canaveral, onto a concrete pad a few miles from where it had lifted off nine minutes earlier. On the live stream the control room in Hawthorne came apart. Engineers who had spent a decade being told the thing was a fantasy stood on their chairs and screamed, and the cheering ran so loud over the broadcast that the hosts gave up trying to talk through it. Two thousand miles east, on the Cape, Elon Musk walked toward the standing booster in the floodlights, a man looking at a machine that had done something machines were not supposed to do. He had been saying for more than a decade that orbital rockets would have to become reusable the way airplanes are, and for most of that decade the claim had been treated inside the industry as the enthusiasm of an outsider who did not grasp why the thing could not be done. He had set the goal in public, attached dates to it, missed several of those dates, and held to the goal across all of it. Now he stood under the lights checking a thing he had insisted on for years against the fact of it standing in front of him. And at a distance from the celebration stood Tom Mueller, who had built the company's first rocket engine by hand in a warehouse in El Segundo when SpaceX was a few dozen people and a lease.

Mueller was a propulsion engineer of the old, obsessive kind, the sort who had built rocket engines for the love of it before anyone paid him to, machining the parts himself and firing homemade motors out in the desert on weekends with the amateurs. The first Merlin had come together in that leased space, one engine, built and torn down and rebuilt by a team small enough that everyone on it could see everyone else across the floor.

By the autumn of 2008 the company had very nearly run out of road. Its first rocket, the small Falcon 1, had flown three times and failed three times, and each failure had burned money the company could not replace. Musk had poured the fortune he made selling his earlier companies into SpaceX and into Tesla at the same time, and by that fall both were close to the bottom of the barrel. The fourth flight of the Falcon 1, at the end of September, finally reached orbit, and it did so with the company facing another failure it could scarcely afford. The nine-engine test came that November, while the survival of the whole enterprise still hung on what it did next, and it was the first hard evidence that the larger rocket the entire plan now depended on had a heart that would hold together when it was lit.

Seven years before the landing, Tom Mueller's propulsion team had brought Merlin to a full nine-engine cluster, the first stage the larger vehicle would depend on. In November 2008 they fired all nine together at McGregor for a mission-length test. Nine plumes met beneath the stand and the Texas scrubland took the noise. Mueller had said the team was elated when the engines first worked together; now the cluster held for the length of a flight. The aerospace trade press had barely noted the test. The company was small, the rocket had not yet flown, and there was no reason for the industry to look. The booster standing on the pad in 2015 was the dividend that test fire had been paying out for seven years, and inside the industry the meaning of the night took a good deal longer to land than the booster had.

A rocket stage had returned from the edge of space and set itself down intact. The cost of a launch is dominated by the cost of building the vehicle, and a vehicle flown more than once spreads that cost across every flight it survives. Whether it could then be flown again cheaply, or safely, or soon, was a question for the mornings after. Coming back whole was the thing the entire industry had agreed could not be done at a price worth paying, and what the landing was finally worth came clear the next morning, when the stage could be rolled into a hangar whole and dry and taken apart by engineers who wanted to fly it again, where every booster before it had come back as wreckage on the seafloor or had never come back at all.

The landing was harder than the cheering quite conveyed. A Falcon 9 booster coming home is nearly empty, its tanks all but drained, and even a single one of its nine engines throttled down as far as it will go makes more thrust than the returning booster weighs. The booster cannot hover. It cannot ease itself down and feel for the ground the way a helicopter does. It has to fall, lighting its engine at the last possible moment and timing the burn so that it arrives at zero speed in the same instant it arrives at zero height, a maneuver the engineers called a hoverslam and the rest of the world would recognize as threading a needle at the bottom of a dive. Too early and the booster halts above the ground and starts climbing again; too late and it flies itself into the concrete. A computer flew the descent every time, since no human reflex was anywhere near fast enough to ride it down.

The next decisive flight came on March 30, 2017. A Falcon 9 lifted the SES-10 satellite on a first stage that had already flown and landed, then brought that stage down again. The landing in 2015 had shown that a rocket could return whole; this launch showed a customer payload riding on one that had returned. The question became how many such flights the same tanks, engines and inspection method could support.

THE REUSE CURVE

Watch a single booster across its life and the curve draws itself. The early reused stages flew twice, which the industry had said was impossible, and then a small number of them flew five times, which the industry had said was reckless. Particular boosters in the fleet passed ten flights, then fifteen, then twenty, each one a tail number with a flight log longer than some entire national rocket programs. The crews came to know each airframe the way a ground crew knows a particular airplane, and the time it took to turn a landed booster around and fly it again fell as the fleet matured, from something like four months in the early days down toward six weeks, and then down again as the crews learned each stage.

The booster was the largest single piece worth recovering, and the company learned to recover more than the booster. The payload fairing, the aerodynamic nose cone that shields the satellite through the thick lower atmosphere and then splits into two halves and falls away once the rocket is in vacuum, cost a great deal to build, and for years every flight had dropped both halves into the sea. The company first tried catching the parachuting halves in giant nets strung above fast boats, sending the boats chasing back and forth across the ocean under the falling hardware, and when the catch proved too chancy it settled for fishing the halves out of the water, drying them, and flying them again. Every piece pulled back from the ocean was a piece that did not have to be built from raw metal for the flight after it.

What fell with it was the price, and the mechanism behind the fall was no stranger than the one that makes a taxi ride cheaper than buying a car to take a single trip. A booster flown once had to carry the entire cost of its own manufacture on that one flight. A booster flown ten times carried a tenth of that cost on each, and the remaining costs of a launch, the propellant and the range fees and the labor of refurbishment, were small enough beside the price of a new vehicle that the cost of a flight began to settle toward the cost of getting the same booster ready to go again. The fifty-four thousand dollars a kilogram of the Shuttle era gave way to figures on an expendable Falcon 9 in the low thousands, something near twenty-seven hundred dollars a kilogram, and lower again on a reused one, closer to fifteen hundred, with the projected cost of a fully reusable Starship at scale reaching down toward a couple of hundred dollars a kilogram or less. Set those numbers in a row and the shape they make is a staircase, each step a thing that had been declared impossible becoming, within a few years, the floor that the next impossible thing was measured against.

The per-kilogram figures were the cleanest way to watch the fall, and the price of a whole launch told the same story in plainer terms. A Falcon 9 came to market listed around sixty million dollars a flight, against much more expensive expendable Western launch services. Reusing the costly first stage meant a mission could carry less of that stage's manufacturing cost each time it flew. A payload team could spend more of its budget on the instrument it wanted to send up, or afford to send another when the first taught it what to change.

For the satellite operators who bought launches, the change arrived as vertigo. Businesses that had spent a decade engineering a single spacecraft to be flawless, because a launch was a once-in-a-program event that could not be risked, now found that reaching orbit was the cheap part of the enterprise, and the old caution that had governed every decision about mass and margin and redundancy began, slowly, to look like a habit carried over from a vanished economy.

The cumulative effect on launch traffic was a new cadence. By 2025 the reusable fleet was lifting more mass to orbit each year than the rest of the world's launch operators combined. The mass included satellites made in batches, crew supplies, scientific instruments and cargo for machines already in orbit. A booster went up, came back and went up again, spreading the cost of its making across more trips. The frequency meant a builder no longer had to treat a missed launch window as the loss of an entire year's chance to test an idea.

A thirtyfold drop in the cost of reaching orbit makes a different set of things worth doing. Constellations of thousands of satellites, where the old cost of orbit had allowed only handfuls. Payloads cheap enough to be designed for replacement, stamped out on an assembly line and flown in bulk and left to burn up when their few years were done. Cargo carried to orbit on a schedule that in any other industry would be called freight. Those falling launch prices measured the size of the door that had just been propped open, and almost nothing that would walk through it afterward could have afforded the trip on the other side.

THE CATCH

On October 13, 2024, at Boca Chica, the company caught its biggest booster with the launch tower for the first time, a recovery nothing like the leg landing that brought a Falcon 9 home. The vehicle was Super Heavy, the first stage of Starship, far larger than any Falcon, taller and heavier than the rockets it had grown out of. On the fifth full test flight of the system it came back from a sub-orbital arc and flew itself down toward the same tower it had launched from. The tower carried two enormous steel arms, hinged and counterweighted, that the crews had nicknamed the chopsticks. The plan, which sounded insane when stated plainly, was to have the descending booster slow itself almost to a hover beside the tower and let the arms close around it just below the grid fins, so that the vehicle never touched the ground at all and could in principle be set straight back onto a launch mount.

The reason to attempt anything so difficult was weight. Landing legs strong enough to hold a booster the size of Super Heavy would themselves weigh many tons, and on a rocket every ton of legs is a ton that cannot be propellant or payload. Catching the booster in the air moved that weight off the vehicle and onto the tower, which never had to leave the ground. The catch also closed the gap between landing and the next launch, because a booster set straight back onto its mount skipped the slow business of being safed, hauled in from a landing site, and stood back up. The whole design pointed at a rocket that could fly, return, and fly again on something closer to the turnaround of an aircraft than of a spacecraft.

The precision the maneuver demanded was of a different order from a landing on a pad. A booster setting down on concrete or on a barge has a few meters of margin in any direction. A booster flying up into a pair of arms hung from a fixed tower has almost none. It has to arrive at a particular point in the air, defined against a structure it must not strike, and hold itself steady enough there for the arms to close on a narrow band of its body while its engines are still burning to keep it aloft. The first time the company committed to the catch on a real flight, a failure stood to cost it both the booster and the tower it had spent a year building. It caught the booster on the first attempt.

Set against the scorched stage riding a barge through the dark some years earlier, the catch was the same idea carried to its end. The first landing had proved that a stage could come back. The catch was an effort to make coming back so fast and so cheap that, for the largest rocket ever flown, the cost of the trip would at last stop deciding what could be done in space.

The booster came down out of the sky faster than sound, lit a subset of its engines for the landing burn, and shed its speed in the last hundred meters in a controlled fall that brought it to a near stop precisely where it needed to be. The tower arms moved inward below the grid fins while the Raptors kept firing, and the catch hardpoints on the booster, load bearing pins built for this contact, settled onto the steel with the vehicle nearly motionless in the air. For a moment the engines still held thrust as the weight transferred into the arms, then the burn ended and the Raptor engines began to tick as they cooled. Super Heavy came to rest intact against the launch tower, the machine that had left the pad under thrust now secured by the hardware waiting to receive it.

At the press site the crowd did the thing crowds do when they have just watched something they did not entirely believe would work. For a beat there was silence, the held breath of several thousand people checking the picture against their own disbelief. Then the sound broke over the flats, the shouting and the disbelieving laughter, the noise of people who build things watching one of the hardest of them finally come true. The booster hung from the tower in the first light, the stainless steel still warm, two steel arms holding a skyscraper out of the air against the flat Texas sky.

3

The Constellation

THE TERMINAL IN THE FOXHOLE

THE BOX WEIGHED ABOUT as much as a loaded toolbag, and a soldier carried it up a frozen embankment in a backpack with the flat antenna strapped across the top. It was the last week of February in 2022, a few days into the invasion, and the position was a scrape of earth above a road somewhere in the east of the country where the cellular network had gone dark, the fiber had been cut, and the only thing still arriving over the old channels was jamming. The soldier had never handled one of these before. He set the panel down in the snow, ran a short cable to a vehicle battery, and let the device do the one thing it had been built to do without instruction, which was to find the sky by itself.

The panel tilted on its motorized base and began to hunt, sweeping the whole dome of the sky for a satellite that was moving fast, one that would rise on one horizon and set on the other within a few minutes, never hanging in a fixed slot above the equator the way the old satellites did. It locked. Soon after a cold start, on a hillside with no tower standing for tens of kilometers and no working ground network of any kind, the terminal had opened a two-way path to the internet fast enough to carry video.

The hardware had reached that hillside through a request made in the open. Mykhailo Fedorov, the country's minister of digital transformation, a man in his early thirties who had spent the previous two years moving Ukraine's government onto smartphones, did not file a procurement order or work a diplomatic back channel. On the twenty-sixth of February he wrote to Musk directly on Twitter, in front of everyone, and asked him to send Starlink terminals. The reply came the same day and ran to two short sentences, saying that Starlink service was now active over Ukraine and that more terminals were on the way. The first pallets crossed the border within days, hauled in on trucks beside the ammunition and the field dressings. Within weeks, terminals were operating across the country. Within a few months the count had climbed into the thousands and kept climbing, the dishes turning up at command posts and field hospitals and artillery positions and in the basements where engineers were fighting to keep the power grid alive.

Fedorov belonged to a government that had spent its peacetime years moving itself onto the phone. The ministry he ran had built a single app that put documents and public services on a phone, a state that lived in a citizen's pocket with the office and the queue dissolved into a screen. A minister who thought of government that way saw nothing strange in conducting foreign procurement on a public timeline in front of millions of onlookers. The openness was the method. A request made in the open was one a company could answer in the open, within hours, skipping the months a defense ministry would have spent drafting a contract, in a week when the towers were already going dark.

The hardware had never been designed for military use, no piece of it built with a war in mind, and it had not needed to be, because the entire product had been engineered around a single assumption, which was that the hardest part of getting a person online should be carrying the box to a place with a clear view of the sky. Everything downstream of that, the pointing, the handoff from one fast-moving satellite to the next, the encryption, the routing back down to a ground station hundreds of kilometers away and out onto the open internet, happened on its own. A consumer product built to spare suburban customers a call to technical support turned out, when it was carried into a country whose communications had been deliberately destroyed, to be the most resilient network in the theater. The capability had been accumulating overhead for three years, and the war put it to the test, in front of the whole world.

What the link carried, once it was up, was everything. Commanders flew reconnaissance drones over it and watched the feeds in real time. Artillery crews took targeting data through it. The engineers fighting to hold the power grid together coordinated their repairs over it while the cellular network around them stayed dead, and hospitals kept their records reachable, and banks kept clearing. A system sold to civilians as a convenience had become, inside a week, the nervous system of a country at war.

The other side tried to take it down. Russian electronic-warfare units, which had spent the opening weeks shutting off Ukrainian radio and cellular links almost at will, turned the same jammers on the terminals, and at first they found purchase, the signal degrading wherever the interference was strongest. What happened next was the thing the old communications gear could never have done. The company pushed new software to the terminals over the air, an overnight phone update sharpening the receivers until they could pick the signal out from under the noise, and the links came back. The jammers adapted and the software adapted again, a contest fought entirely in code and run from a console thousands of kilometers from the front. A network that could be reprogrammed faster than it could be jammed was something new on a battlefield, and it held because the people who built it could change it nightly without anyone going near the position.

That dependence ran in both directions, and the second direction took longer to surface. A war now leaned on a network owned by a single private company, and over the months that followed the question of where the service worked, and whether it worked, became its own quiet front. The terminals had arrived as a gift and a rescue. They were also a piece of national infrastructure that sat, for the first time in the history of war, outside the control of any nation fighting with it.

SpaceX developed Starshield for government and national-security use, drawing on the satellites, launch system and laser links it had built for Starlink. It offered a more secure network and equipment for government customers. That answered some demands the consumer service could not, but it did not by itself make the network state-owned. A country relying on a privately built link still had to settle what it could command, what it could inspect and what would happen if service changed during a crisis.

WHAT'S UP THERE

The thing the terminal was talking to had become a moving structure of more than ten thousand working Starlink satellites by autumn 2026. More rose with successive Falcon 9 launches, dispensing off the top of the second stage like a deck of cards fanned into the dark. Each satellite was cheap where it counted, cheap enough that the network could absorb the loss of one while its neighbors took the traffic. Much of the constellation flew in low Earth orbit near five hundred and fifty kilometers. Atmospheric drag there would eventually bring a failed satellite down, while the short trip to orbit and back kept signal delay low. Neither feature removed the work of collision avoidance or the responsibility to keep debris from accumulating in the paths the next spacecraft would need.

The ten thousand flew in shells, sets of orbital planes tilted at fixed angles to the equator and spaced so their paths wove a moving lattice that covered the populated latitudes without a gap, then a second shell at a steeper tilt to reach the high latitudes the first one skimmed, and a lower shell still being filled in to thicken the coverage over the busiest ground. Within each plane the satellites flew in an evenly spaced string, each one trailing the last like cars on a circular track, so that as one slid out of a terminal's view the next had already climbed over the opposite horizon. Coverage belonged to the whole formation, managed from the ground so that somewhere overhead, always, one satellite was rising to take the handoff. The largest structure humans had ever placed above their heads was a pattern of motion, continuous and never still, that covered the planet by refusing to hold any position at all.

Cheap at that scale was a manufacturing achievement before it was an orbital one. Spacecraft had always been handmade objects, assembled a few at a time in clean rooms by technicians in white suits, each one a bespoke machine. To fill the sky with thousands of them the company built a production line in Redmond, Washington, that was turning out about seventy satellites a week. The satellites came off that line flat, packed by design so that two dozen would stack inside a single fairing and fan off the top of the upper stage at the drop-off point. Each one then raised itself the rest of the way, climbing slowly from the altitude where it was released up to its operational shell on a small onboard thruster over a span of weeks, finding its assigned place in the lattice with no hand touching it.

Keeping Starlink running meant building and launching replacement satellites year after year. With satellites designed for about five years of service and aging spacecraft deliberately brought down, the company had to keep launching simply to stand still, putting up new satellites at the same pace the old ones burned away. A fleet that had to be rebuilt continuously could also be upgraded continuously, because each fresh batch coming off the line could carry better antennas and faster processors and more capable lasers than the batch it replaced. The sky filled, over the years, with successive generations flying together, the newest satellites more powerful than the ones still finishing their service overhead. A traditional satellite was frozen at the moment it launched, carrying whatever had been current the day it was built for the twenty years it would hang in its slot. This one improved while it flew, version over version, the old hardware retiring itself on schedule to make room for something better.

The satellites that had carried international communication for half a century sat in geostationary orbit at thirty-six thousand kilometers, parked above a fixed point on the equator, each one a bus-sized machine that cost in the hundreds of millions of dollars and took years to build and insure and fly. They had to be that expensive. There were only a handful of them, and a single failure was a catastrophe measured against a quarter of the planet's bandwidth, so each one was built like a cathedral, gold-plated and tripleredundant and never to be touched again once it reached its slot. The whole architecture was an answer to the cost of getting to orbit. When a kilogram going up cost what a small car cost, you flew few satellites and you made each one perfect. Drop that price by an order of magnitude and a different answer becomes available, which is to fly thousands of cheap ones and let the network as a whole be the reliable thing while no single piece of it has to be.

For all that the constellation would end up doing, the thing it was built to sell was ordinary. The customer was a household at the end of a dirt road that the cable companies had never found worth wiring, a farm, a fishing boat, an airliner that wanted to offer its passengers a connection over the middle of the ocean, a construction site, a disaster zone where the towers had washed away. A flat panel on a roof or a deck, a subscription billed by the month, and a connection that behaved like the one in a city apartment. The network carried twelve million unique service lines. Most users had never carried a dish up a frozen embankment or aimed it at anything more dramatic than a call home. The war was one test of the capability. Its daily work was on rooftops, boats and roads.

The resilience the soldiers leaned on lived in the swarm itself, in the sheer number of satellites moving so fast that the loss of any one was invisible to the person on the ground, whose terminal had already handed off to the next satellite rising over the horizon before the first had set. Reliability had been moved out of the hardware and into the swarm.

The ground network the swarm answers is more concentrated than it looks. Close to all of the world's international traffic still runs through undersea cables, and a striking share of it funnels through a single corridor across Egypt, mended when it breaks by a fleet of only around sixty cable ships working a hundred-odd faults a year. A planet's communications hang on a few cuttable threads at a few physical places. That is the vulnerability a self-healing swarm overhead is built to remove.

The satellites also talked to one another. Each carried lasers that linked it to its neighbors in orbit, so that a packet of data could cross the constellation from satellite to satellite through open space and come down only when it was over a ground station near where it was headed. The network reached, that way, into places no ground station could ever see, the middle of the oceans, the polar routes the airliners fly, the empty quarters of the world where no company would build a relay on the ground. The product was a mesh hung in the sky, each satellite handing data to the next through vacuum, touching the dirt only where a paying customer happened to be standing.

Reaching a place physically and being allowed to sell there were two different things. Radio crosses a border without asking. The right to broadcast into a country does not, and every government that wanted the service inside its territory had to license the company to use the airwaves over its own citizens. The constellation had to be negotiated into existence one country at a time, a patchwork of market-access deals and spectrum grants and local ground stations, until the map of where Starlink worked became a map of where governments had said yes. Some said no, unwilling to let a foreign company own the pipe their citizens spoke through. Others said yes and later wondered what they had agreed to, because a network that reached every rooftop was also one no national regulator could fully switch off once the dishes were in people's hands. The hardware flew over everyone. The permission to use it was something each nation handed over, or withheld, one negotiation at a time.

Ten thousand objects moving at orbital speed is also a responsibility, and the company carried it in software. Each satellite tracked the catalog of other things sharing its altitude and steered itself clear when a close approach loomed, tens of thousands of small automated dodges a year that no ground controller could have flown by hand. The fleet drew complaints of another kind from astronomers, whose long-exposure images began to fill with the bright streaks of satellites crossing the frame, and the company answered by darkening the later spacecraft with coatings and sunshades to dim them against the sky. Filling the heavens with hardware turned out to be a thing with a bill attached, paid partly in vigilance and partly in apology.

DIRECT TO PHONE

The next gap to close was the one between the dish and the phone in a person's pocket, and in 2025 the company closed it. Through 2025, in partnership with T-Mobile, Starlink began offering a service that did something the industry had spent decades assuming was impossible. A standard smartphone, the ordinary slab in an ordinary pocket, with no special antenna and no modification of any kind, could send a text message straight up to a satellite passing a few hundred kilometers overhead. Mike Sievert, who ran T-Mobile, framed it around a simple promise, that dead zones could shrink. And the dead zones were not small. Cellular coverage maps, the green blobs that carriers print over their territories, cover the places where people are dense enough to justify the cost of a tower. They leave most of the physical surface of the planet uncovered, the deserts and the mountains and the open ocean and the long empty stretches of highway where a broken-down car had always meant a walk. A constellation that talked to phones directly did not care where the towers were, because it did not need towers. The ground infrastructure it depended on was the one already in the customer's hand.

The reason the industry had called this impossible was a matter of physics that had not changed. A phone is a weak radio. It pushes out a fraction of a watt through an antenna a few centimeters long, barely enough to reach a tower a mile or two off, and a satellite a few hundred kilometers up is hundreds of times farther away than the farthest tower. For the same phone to close that link, the work had to be done at the other end. The satellites built for the service carried unusually large antennas, phased arrays several meters across, sensitive enough to hear a phone whispering up from the ground and powerful enough to be heard whispering back. The cleverness was in refusing to touch the phone at all. Hundreds of millions of handsets already in pockets became, with no new hardware and no instruction to their owners, the ground terminals of a satellite network, because the satellite had been built to meet an ordinary phone exactly where it already was.

There was a second trick beside the antenna, and it was a legal one. A phone only knows how to talk on the frequencies its carrier licenses, the specific slices of radio that a company like T-Mobile had bought the rights to use across its ground network. So the satellite spoke the same language, broadcasting on T-Mobile's own terrestrial frequencies from orbit and presenting itself to the phone as one more cell tower that happened to sit a few hundred kilometers up. The handset needed no new chip and no setting changed, because as far as it could tell it had simply found a tower where there had never been one. Making that work meant convincing regulators to let a licensed ground frequency be beamed down from space over the same territory, a question no rulebook had been written to answer, and the answer came country by country in the same slow patchwork that governed the dishes.

The service started with text, because a single message is small enough to squeeze through a link that thin. Voice and slow data followed, and faster data would come as the newer and larger satellites went up. Each step widened what a phone in a dead zone could do, from sending word that you were alive, to holding a conversation, to pulling up a map of the way out. The carriers that signed on early stopped advertising the reach of their tower network and started advertising that their service simply worked everywhere, which promised something the old tower map never could.

Other builders tried a different geometry. AST SpaceMobile designed large unfolding antennas to make a direct link with ordinary phones, while Starlink adapted a large fleet already moving overhead. Both approaches placed the difficult radio work in space so the phone could remain the phone in a person's pocket. The first useful messages were small. A text from a valley without a tower could tell someone where to come, while wider service would need more satellites, capacity and permission to use ground-carrier frequencies.

There is a particular discomfort that settles over an industry when the expensive thing it built its position on turns out to be answerable from an angle its towers could never defend. The cellular carriers had not lost. Their networks still carried nearly all of the world's mobile traffic and would for years. But the question had changed under them, from how much it costs to cover the next valley with a tower, to whether the valley needs a tower at all.

THE STANDING RESOURCE

On the twelfth full flight of Starship, in May of 2026, two modified Starlink satellites went up with an unusual assignment and an internal nickname, the Dodger dogs. As the pair deployed, their cameras looked back at the ship and its heat shield from a vantage no camera mounted on Starship could give. The images traveled through the same communications architecture built to carry service across the Earth. Engineers could inspect the skin of their spacecraft from another machine falling away from it, a new angle on a vehicle whose return would again put its tiles through violent heating.

A consumer internet service, sold by the month to households who wanted Netflix in places the cable companies had skipped, had given the rocket builders a second vantage on a spacecraft they were still learning to bring home. The same factory that turned out relays for rural broadband could turn out, with small changes, a pair of camera-bearing spacecraft able to look back at the ship after deployment. The constellation had been built to move bits for paying customers. Once it existed, at that scale and that price, it became raw capacity that the people who owned it could point at any problem they happened to have, including the problem of building the next rocket. A thing you have already paid to put in orbit by the thousand stops being a product line and starts being a standing resource, available for uses nobody wrote into the business plan.

The reentry cameras were the visible case. There were quieter ones. Researchers had begun working out how to navigate by the constellation itself, reading the precise timing of its signals to fix a position on the ground the way the global positioning satellites do, a backup that mattered most in exactly the places where the positioning signals were being jammed. A fleet built to sell broadband carried, as a side effect of knowing precisely where each of its thousands of members sat at every instant, a second usefulness as a clock and a map in the sky. It fell out of the scale.

CHEAP, REPLACEABLE, SCALED

Once that was clear, the same realization began to spread, and people started asking different questions about space. For fifty years the governing question of any space project had been narrow and defensive. What is the satellite for, the financiers would ask, meaning which single expensive mission would justify the enormous fixed cost of the launch, because the launch was the thing that ate the budget and the satellite had to earn it back alone. When the launch stopped eating the budget, the question opened up. People stopped asking what a satellite was for and started asking what a constellation was for, the question you ask about a piece of infrastructure. Roads are not for one trip. A constellation, once it is cheap enough to be flown by the thousand, is not for one mission either.

So the sketches started beside working fleets. Companies proposed orbital data centers, with solar arrays feeding processors and radiators sized to carry their waste heat away as infrared; how much power the arrays made and how much heat the panels could shed would set the useful load. Others worked on refueling depots, tanks parked in orbit so that a vehicle could top off after reaching space and press on farther. Earth observation had already crossed into commercial work. Planet operated imaging constellations that returned a near-daily view of Earth's land, while its higher-resolution satellites could be tasked for another look as their orbits brought them over a place. More satellites and cheaper replacements could shorten that wait. The running imaging service showed one kind of infrastructure a repeatable route to orbit could sustain, and left open what else builders might put there.

The ideas themselves were old. The 1990s had seen a first wave of low-orbit ambition, companies that proposed exactly this, fleets of small satellites ringing the Earth to carry calls and data to anywhere on the surface. They built real hardware and launched some of it. Then almost all of them went broke, and they went broke for one reason above the rest. Putting the fleet up cost more than the service could ever earn back, because every kilogram still rode to orbit on an expendable rocket at the price of a house. Iridium, the most famous of them, flew its constellation, switched it on, and filed for bankruptcy in 1999 within months of opening for business. The engineering had worked. The economics had killed it, and the economics came down to the price of the ride. What the second wave had that the first one lacked was the rocket that flew itself home.

Iridium survived the first low-orbit crash and found durable work in links for ships, aircraft and remote sensors. Its narrower network kept doing jobs that demanded little bandwidth and high dependability. The larger constellation opened other jobs because it could replace satellites frequently and carry more traffic. Both histories show why a route to orbit matters. A working spacecraft can be excellent at its task, yet the price and cadence of launching the next one decide how many different tasks builders dare to give the sky.

THE NEW FLOOR

Once rockets returned and satellites multiplied, the old accounting gave way. Low orbit became a stocked shelf, with bandwidth, cameras, timing, and mass overhead already paid for and waiting to be used. What had been a destination became the place new work started.

PART II

THE INDUSTRIAL BASE

4

The Energy Ceiling

THE DATA CENTER THAT ATE THE GRID

ON THE SOUTHERN EDGE of Memphis, inside the shell of a shuttered appliance factory, a machine came together in 2024 that the crews fitting it out could not fully explain to their neighbors. They were not building from the ground up. The hall already stood, a long windowless building floored with concrete, and the speed that made the project notorious came from skipping everything a new structure would have demanded, the years of foundation and steel and permitting, and going straight to filling the shell that was already there. It was a shed in the loose sense that an aircraft carrier is a boat, its cabinets packed so densely that the engineers spoke of the heat in the same breath as the computing. Inside would sit the largest single concentration of artificial-intelligence processors yet assembled in one place, a machine its owners called Colossus, a hundred thousand graphics chips wired into a single trainable brain and then, within months, as many again.

What made the building notable inside the industry was the speed as much as the size. The first phase had gone from empty hall to a running cluster in a matter of months, a timeline that data-center veterans called impossible right up until they toured it, because the usual job of standing up a facility at that scale ran for years of design review and procurement. Nvidia's Jensen Huang, whose company supplied the processors everyone else was waiting for, later called the build superhuman and said only Elon could have done it. The crews worked the site around the clock, fitting out the hall in sections and switching on the early rows of machines while the later rows were still being bolted down.

Inside, the density was the point. The cabinets stood in long rows, each one packed with more processors and drawing more power than any server rack had been designed to hold, and every chip had to be wired so it could talk to every other chip at once, because a training run splits one enormous calculation across the whole machine and stalls if any part falls behind. The cabling that carried that conversation ran for thousands of miles inside the single building, a web of fiber and copper threaded between the racks. A cluster like this works as one computer the size of a warehouse, fed and cooled and driven as a single object.

The chips could be bought, in volume, from a single company in California whose order book the whole industry was fighting over. Power was the part that could not be ordered off a shelf. At full build-out, the wider campus was projected to draw on the order of a gigawatt, the output of a mid-sized nuclear plant. Feeding that much computing capacity would require power around the clock, even as synchronized training jobs made the draw rise and fall over shorter intervals. A gigawatt is a hard number to feel. It is roughly the steady draw of a city of a million people, or the entire output of one large reactor, potentially devoted to an expanded campus of chips on one industrial site. The appliance factory that had once filled this building drew a small fraction of that. At the proposed scale, the campus would demand power comparable to a city beside it. The local utility, asked to deliver that much power to a single customer on one parcel, did what utilities do. It put the request in a queue. The transmission lines and substations that would carry a gigawatt to one address had to be planned, permitted, financed, and built, and the honest estimate for all of that ran to years, on some readings the better part of a decade.

Memphis had merely joined a national queue. Across the country interconnection requests for generating plants and storage projects had swollen into a backlog measured in years. Their proposed capacity, much of it solar, wind and batteries, exceeded the existing generating fleet. Data centers entered from the demand side, seeking power while prospective suppliers and the wires to reach them waited on studies, upgrades and crews. Building a wire, it turned out, had become slower and harder than building almost anything the wire was meant to connect. A company that accepted its place in that line was accepting a schedule set by congestion and paperwork and the physical labor of stringing high-voltage cable across contested ground. The wait was not only a matter of permits. The grid itself was assembled from parts that had become hard to get. The large power transformers that step current up and down between a generating plant and a city are custom machines the size of a house, most of them built overseas, and the orders for them had backed up until the lead time stretched past two years and kept climbing. High-voltage switchgear, the specialized electrical steel, the crews trained to string transmission line, all of it was scarce at once, because the whole country had resolved to build at the same moment after decades of building very little. A company that wanted power in two years could not buy its way past a transformer that took three to arrive. The ones in a hurry stopped accepting it.

A decade is not a timeline a frontier AI company recognizes. The model being trained inside the shed would be obsolete long before the substation was energized. So the operator gave the utility's answer the only reply that kept the schedule alive. It brought the power plant to the site. Through 2024 and into 2025, flatbed trucks delivered more than two dozen mobile natural-gas turbines, the kind normally wheeled in to cover a summer demand spike for a few weeks, and parked them in a row beside the building, and ran them continuously. They generated several hundred megawatts on private property, behind the meter, while the grid connection crept forward in its queue. The arrangement drew objections from people who lived downwind and questions from the agencies that issue air permits. When a company needed a gigawatt and the grid offered it a place in line, the company found that it was faster to become a power producer than to wait to become a customer.

To stand near the site once the turbines were running was to hear the new economics before anyone explained it. A bank of mobile turbines on flatbeds, burning gas day and night, threw a low continuous roar across a flat parcel that a few years earlier had been the quiet ground of a shuttered appliance plant. Tanker trucks came and went. Transformers the size of houses sat in fenced yards. The old footprint had been swallowed by a structure with no windows and no obvious purpose to a passerby, and the only clue to what it did was the sheer quantity of energy visibly pouring into it. A neighborhood that had lost its factory had gained the densest concentration of computing power on Earth, and what the two had in common, across the decades between them, was that each had been placed on that ground by a hard calculation about power.

Power going in becomes heat coming off, almost all of it. At the projected gigawatt scale, the wider campus would have to throw roughly a gigawatt of heat back out before its processors cooked. The early AI halls had tried to do this with air, vast fans moving rivers of it past the chips, and the chips had outrun the air. The Memphis machine used liquid, water and coolant pumped directly across the hot silicon and out to banks of chillers, which meant the site needed not only power but a matching flow of cooling water, drawn and circulated and given somewhere to go. The heat was a second utility on the same scale as the first, planned and plumbed into the site before a single chip switched on.

The water that carried the heat was a third. A campus shedding heat at that projected scale would need an enormous cooling system, and an evaporative design would need a steady water supply drawn from local rivers or aquifers or municipal mains, warmed as it carried the heat and then returned or evaporated away. In a wet region this is a manageable line on the budget. In the dry stretches of the country where land and power ran cheap and the new campuses were rising fastest, water was already spoken for, and a data center asking a small county for millions of gallons a day put itself in line behind the farms and towns that had drawn from the same source for a century.

For an operation of this kind in the middle of the 2020s, electricity had stopped being a utility bill and become the first line on the engineering drawing. The power supply was sketched onto the site plan before the racks, sized before the cooling, argued over before the software. The turbine sitting outside the building, humming, burning gas, decided how large the machine could grow, because the rate at which energy could be delivered to that square of ground was the rate at which intelligence could be manufactured on it.

THE OLDEST CONSTRAINT

Stand in front of that building long enough and the particular case opens into a general one, because the constraint pressing on Colossus is the same constraint pressing on every ambition that wants to get physically larger.

Everything that scales is gated by the rate at which energy can be produced, moved, and stored at industrial scale. A training run is gated by it. So is a factory, which is a building for converting electricity and raw material into objects at a certain pace, and cannot exceed the pace at which the current arrives. So is transport, and mining, and the desalination of seawater, and the smelting of metal, and every operation that anyone has ever proposed to carry out beyond the surface of this planet. The common unit underneath all of them is joules per second, the watt, the rate of energy flow, and there is no trick of cleverness that converts a small flow into a large outcome. Chips and algorithms and software arrange themselves around the available power and wring more from each watt without ever making one. All of them sit downstream of the energy and inherit its ceiling.

Storage is the leg of that triad people forget, and it is the one that decides whether the other two are worth anything. Energy has to be available at the moment it is needed, and the cheapest ways of producing it have a habit of producing it at the wrong moment. The sun delivers its enormous flux at midday and nothing at night. The wind blows hard one day and falls still the next. A grid built on those sources makes a surplus no one can use at noon and a shortfall no one can cover at dusk, and a training run or a smelter or an assembly line cannot pause for the weather. Between the moment energy is generated and the moment it is consumed sits the problem of holding it, and a society that cannot store energy cheaply is forced to keep burning fuel it can switch on and off at will, which puts a floor under how clean and how cheap its power can ever be. Whoever can store energy at the scale at which it is produced changes what the whole system is allowed to run on.

That timing problem shows up in dispatch. Operators curtail wind and solar farms while wind and sunlight are available, then bring gas plants on to cover the evening peak. A unit stored from the surplus is a unit of clean power not wasted and fuel not burned. The ability to carry that energy across the hours helps set how cleanly and cheaply a grid can run.

Efficiency is the move everyone reaches for first, and it works right up until it doesn't. A better chip computes more per watt, a better engine drives more per liter, a better furnace smelts more per ton of fuel, and each improvement looks for a moment like a way around the ceiling. What history shows is that efficiency lowers the cost of using energy, and lowering the cost of a thing causes people to use far more of it. The efficient engine, by making power cheaper to put to work, ended up in everything and drove total fuel consumption up. The efficient chip made computation cheap enough that the industry found a billion new uses for it and then built a machine in Memphis that eats a gigawatt. Efficiency raises the ceiling. It never removes it, because the gains are spent on doing more, and doing more runs on power.

There is a reason the answer to the ceiling has always been a change of substrate. Cleverness conjures no new energy, so the only way a society has ever lifted its ceiling is by moving to a source that packs more energy into less mass, or one that can be moved and switched and metered with more precision. A denser, more controllable source lets a builder concentrate more power in one place and aim it at one task, and concentration is what every large undertaking needs. The history of what people have been able to build is, underneath the inventions and the names, a history of the substrate growing denser and more biddable, each new one unlocking a class of things the previous one could never feed.

This is the oldest constraint there is, and the history of expanding human capability can be read as a sequence of answers to it. Each major step up in what people could physically do has been a step onto a denser energy substrate. Wood gave way to coal, and coal let a society run engines that no quantity of firewood could have fed. Coal gave way to oil, which carried more energy per kilogram and could be poured, and the century of cheap oil built the highways and the airliners and the chemical industry. Oil shared the stage with the electrical grid, which let energy generated in one place be used instantly in another, and electricity is the substrate the present runs on. Whatever the next dense substrate turns out to be, the pattern of the transitions is consistent. The civilization that reaches the next rung first does the new things first, and the thing that holds everyone at the current rung is the energy bottleneck, old and patient, the ceiling that every ambition eventually rises to meet.

THE WATT MATH OF INTELLIGENCE

The reason this old ceiling came back into view so suddenly is that one industry started climbing toward it at a rate nobody's grid had been built to expect. For close to a decade, the amount of computation poured into training the largest AI models had been roughly doubling every several months, a pace steeper than the famous doubling of transistors that defined the previous era of electronics. Computation of that kind is bought with electricity. A frontier training run in the middle of the 2020s consumed, over its weeks of operation, an amount of energy that a decade earlier would have been an unusual annual figure for a small town, and the model that resulted was itself expensive to operate, because every query a user typed had to be answered by running the trained network forward, multiplied across hundreds of millions of users asking constantly. The cost of building the model was a spike. The cost of running it was a permanent new load on the grid that did not go away.

For most of a working lifetime, the people who plan the American grid had forecast almost no growth at all. Electricity demand in the country had been close to flat for the better part of two decades, held down by efficient lighting and appliances even as the economy grew, and the utilities had built their plans around a future that looked much like the present. The data centers broke that assumption in the space of a couple of years. Forecasters who had penciled in fractions of a percent began revising their numbers upward by amounts they had not used since the postwar decades, and the revisions kept arriving, each one larger than the last, as another gigawatt-scale facility was announced. A whole profession of long-range planners, trained to expect a calm and predictable load, found itself staring at the steepest demand curve any of them had worked with, drawn almost entirely by machines that think.

Models doing sustained work added to the continuing demand. Drafting, checking and revising across long stretches kept computation running while no person waited at the keyboard. Each user's draw could grow from a flicker toward a steady load.

A human brain runs vision and movement and memory and language together on about the power of a dim light bulb, some twenty watts. Put one hundred thousand H100 GPUs in racks and their rated power alone can reach seventy megawatts. The fans, networking gear and cooling system demand more. Keeping that cluster running takes substations, transformers and cooling towers.

The loads were beginning to change what other people could ask a computer to do. A laboratory could train a larger model, then serve it to people working through designs, code, images or a difficult search. Every one of those uses drew on chips that had to be powered and cooled when the question arrived. A rack whose supply failed halfway through a run could waste hours of work across the machines linked to it. At Memphis, buying processors was only one part of making computation available; the wiring, turbines, cooling loops and people keeping them running made the chips useful.

The powered floor was a physical resource a builder could share with others. A company using a cluster need not own the building to ask its chips to work, but someone had to raise the shell, bring in transformers, give the heat a path out and keep the machines on through the run. That division could let a smaller laboratory work at a scale it could not have built for itself. It also made the laboratory dependent on the people who controlled the site and the current. The question was how many such floors could be prepared, and how quickly, without asking every new user of computation to build a power plant first.

The scramble for that power reached places no one had expected. Technology companies, which a decade earlier had talked about energy mostly in the language of carbon pledges, began signing the longest and largest electricity contracts of their lives, buying the entire output of power plants for years at a stretch to guarantee that the current would be there. One of them struck a deal to restart a nuclear reactor that had been shut down and left for dead, paying to bring a sealed plant back to life to power a fleet of data centers. Others queued for gas turbines whose manufacturers were suddenly years behind on orders, or drew up plans for plants of their own, or chased the long-promised small reactors that no one had yet built at scale. The common thread was a buyer willing to do almost anything to lock down firm power, which flows whether or not the wind blows, because an idle cluster of chips worth billions of dollars is the most expensive thing in the world to leave switched off.

Power being the scarce thing, the map of where these machines got built was redrawn around it. An earlier generation of data centers had clustered near people and near fiber, in the suburbs of the big coastal cities where the customers and the networks were. The new ones went where the electricity was, or where it could be made. They rose next to existing power plants, in stretches of Texas with spare generation and loose permitting, in the river valleys and the open middle of the country where land was cheap and a substation could go up without a decade of objection. A company choosing a site now read the local grid the way an earlier industrialist had read a coalfield or a fall of water, looking for the place where the energy already was, because moving the computing to the power had become far easier than moving the power to the computing.

The same problem appeared wherever builders wanted to put dense computing. A country with abundant generation could offer more sites, but each site still needed wires, transformers, cooling water and trained crews. A country short of those things could have brilliant programmers and still wait years to give their machines a place to run. Increasing the supply meant building more than power plants. It meant building the routes by which power could reach the cluster at the rate its chips required.

The source of the current mattered as much as its quantity. A reactor, gas turbine, solar farm and battery bank can all feed a line, but they leave different costs with the place around it. The temporary turbines at Memphis brought computation online quickly and also brought exhaust, noise and air-permit disputes to its neighbors. Solar could supply abundant daytime power but needed storage or another firm source when the sun went down. A site that planned only for the next rack might leave those consequences to the community and the grid; one built for years had to make its power and cooling answerable there too.

At national scale, data centers were becoming a load that grid planners had to place beside homes, trains, factories and heat. A substation sized for one district could not be promised twice. A new hall of chips might be finished before the transformer and high-voltage connection that could feed it. The question for a builder was no longer only how fast the racks could be filled, but what source, storage and connection would let them run when the first training job began.

In Ireland the data centers already draw something like a fifth of all the electricity the country meters. A small grid tips first, and the larger ones are walking toward the same line.

So the language of the industry changed to match the constraint. The data centers being planned for the second half of the 2020s were no longer described the way the previous generation had been, in square feet and cabinet counts and rack units. They were described in gigawatts, the same unit used for power stations, because that is what they were becoming. A company announcing a new facility now led with the figure for its power draw, and everyone in the room understood that the number was the announcement. The watt had become the measure of the ambition.

THE ROUGE

Henry Ford would have recognized all of it. He ran into the same wall a hundred years earlier and answered it the same way. By the time Ford had a car that ordinary people could afford and a method for building it fast, the engineering of the automobile had stopped limiting him. What limited him was everything that had to flow into the factory and everything that had to be done to it once it arrived. A moving assembly line is a machine for pace, and a line runs only as fast as it is fed, with steel and glass and rubber and, underneath all of it, power to turn the shafts and the motors. The American electrical grid of the early twentieth century was young and patchy, built outward city by city, and it had not been laid down with the assumption that a single manufacturer might want to draw the output of a power station through one set of gates.

There was a deeper reason Ford needed electricity in particular, and it ran back to the line itself. The moving assembly line he had perfected at Highland Park a few years earlier worked because the machines along it ran on their own electric motors, started and stopped and tuned on their own, where the factories of the previous century had run every machine off a single central steam engine through a forest of overhead shafts and leather belts. Electricity was what let the machinery be arranged for the flow of the work, freed from the reach of a single belt. The whole method depended on current being everywhere on the floor and always there, and a method that leaned on power that completely could not be left at the mercy of a young grid that might or might not deliver.

Ford's answer was to stop depending on anyone else's substrate and build his own. At the enormous site on the River Rouge, south of Detroit, he assembled something the world had not quite seen before, a factory that made its own inputs where other carmakers bought theirs. Ore and coal and rubber came in at one end on his own ships and his own rail. Inside the fence stood his own blast furnaces making his own steel, his own glassworks, his own power plant generating the electricity that ran the whole machine on a current he controlled and never had to wait for. The Rouge has been remembered as the high cathedral of mass production, and it was, but the reason it had to be built that way is the reason that matters here. The surrounding industrial economy could not yet supply, reliably and at the scale and on the schedule he demanded, the things his ambition required. So he brought them inside the fence, and the power plant came inside first, because nothing else ran without it.

The scale of the place is hard to hold in the mind even now. At its peak the Rouge employed something on the order of a hundred thousand people on a single site, took in raw ore at one gate, and turned out finished automobiles at another in a span of days, the whole chain of smelting and casting and stamping and assembly compressed onto one stretch of riverbank. Ford pushed the ownership of his inputs to its limit and past it. He bought iron mines in the north and the ore ships to carry their output down the lakes. He bought timberland and a railroad. He went so far as to carve a rubber plantation out of the Brazilian rainforest, a town he called Fordlandia, meant to free his tires from the dealers who set the price of rubber. The plantation failed, defeated by blight and distance and a jungle that would not be run like a production line, and the failure is worth remembering because it marks the edge of the strategy. Bringing the world inside the fence works only as far as a company can actually master what it has enclosed. The power plant, the steel, the glass, those Ford could run. The rainforest he could not.

What Ford had bought with the Rouge was not only steel and current but tempo. A line fed by its own plant could be arranged around a steady supply rather than stop while another mill or power station caught up. The computing builders in Memphis faced a version of that scheduling problem. A transformer that took years to arrive could hold a finished hall idle; turbines on site could let the first racks run sooner. The power source changed the clock of the work.

A century apart, workers at the Rouge and at Memphis needed a dependable flow of current before the next machine could be useful. Ford built a plant into the factory; the

Memphis team brought temporary generation alongside a repurposed hall. One source was meant to feed a production complex for decades, the other to start racks while a grid connection was still being built. In both, energy supply became part of the machine's design, shaped to the rate of the work inside it.

Ford's failed rubber plantation showed how far that method could be pushed before distance and living systems defeated it. The Memphis halls had limits of their own. The chips arrived through a worldwide chain of designers, foundries and equipment makers that no gas turbine beside the building could replace. A hall could solve one wait for electricity and still wait on silicon. Every large machine was an arrangement of such dependencies; the work was to know which one was holding the next useful step.

THE GATE

Underneath the rack, press, train and desalination pump sits the rate at which energy can be produced, stored and moved. Raise that rate and more of those machines can work at once, in more places, for longer stretches. The gain becomes useful when a crew can count on the current arriving at the next shift, or a town can hold daylight power through the evening. The ceiling moves through wires, storage, generation and people capable of keeping them sound.

THE DESERT PLANT

At Tesla's Nevada battery plant, the next part of that work looked like ordinary industrial motion. Cells came off the line by the millions, storage hardware moved from plan to production, and another kind of capacity was being built as something that could be poured, wired and tested. Power could be made storable at factory scale.

5

Solar, Storage, and the Megapack

THE MILLISECOND RESPONSE

ON 14 DECEMBER 2017, LOY Yang A Unit 3 in Victoria tripped, taking 560 megawatts of generation off Australia's interconnected grid. Roughly 900 kilometers away in South Australia, the loss reached Hornsdale as a falling frequency.

On a fenced paddock outside the town of Jamestown, a row of white Tesla battery cabinets had been standing for only a few weeks, a battery rated to discharge at a hundred megawatts, built on a public dare and finished inside the hundred days its maker had promised or the thing would be free. As the frequency fell, its power electronics answered in milliseconds. Hornsdale sent about 7.3 megawatts into the network while the coal unit was still dropping away. The trace bent under that first injection of power.

On a grid operator's display, the Loy Yang trip would appear as a frequency trace starting down. The battery's response belongs on the same trace, a small, fast addition of power while the much larger coal unit was still losing output. A human operator could read those changing lines, but the first move had already been made by power electronics. Other generators then had time to add their own output. The battery acted in the narrow interval before slower machines answered. Stored energy had become a hand quick enough to touch the grid's frequency as it moved.

The cabinets had been standing on that paddock because of a different failure a year before. In the spring of 2016 a storm had knocked out transmission towers across South Australia and dropped the entire state into a blackout, every light from Adelaide outward gone at once, and the event had turned the reliability of a renewable-heavy grid into a national argument. The state had been closing coal and leaning hard on wind, and its critics said the blackout proved the strategy reckless. Into that argument, in the middle of 2017, came an exchange that read at first like a stunt. An Australian software billionaire asked on Twitter whether Tesla could really fix the problem with batteries. Musk answered within hours, in public, with a number and a deadline. Tesla would install a hundred megawatts of storage within a hundred days of the contract being signed, and if it missed the deadline the installation would be free.

The bet was theatrical and the stakes were small against Tesla's size, and underneath it was a demonstration aimed at every grid operator watching. The cabinets went in with weeks to spare.

The big spinning generators of a power system store energy in their own rotating mass. That inertia begins resisting a sudden change in frequency immediately, buying time while governors and reserve units alter their output. Wind and solar plants often connect through inverters rather than large synchronized rotors, so a grid with fewer spinning machines can see frequency move faster after a trip. Hornsdale brought another response alongside inertia. Its controls sensed the change and sent stored power into the line within milliseconds, before a conventional generator could substantially raise its output. It showed that an electronic resource could take part in the first seconds of a fault, precisely where a changing grid needs a faster hand.

In the months that followed, the paddock outside Jamestown did something the headlines had not predicted. It made money. By answering small frequency wobbles faster and cheaper than the gas plants that had been paid to keep spinning in reserve, the battery began earning a steady income in the market for grid stability, taking a large share of those payments while cutting what the whole system spent on them. A device sold as an insurance policy against blackouts had turned out to be a business. Other operators read the same numbers, and within a few years batteries were going onto grids on every continent, no public dare required.

THE ECONOMICS UNDERNEATH

The speed was the part that made the news. The part that mattered for the rest of the decade was the price, and the price had been falling for years along a curve that almost nobody outside the industry had been watching.

In 2010 a kilowatt-hour of lithium-ion storage cost on the order of twelve hundred dollars, which put grid-scale batteries in the category of things that were technically possible and economically absurd. By the time the Jamestown cabinets went in, the figure had fallen by most of the way to a few hundred dollars, enough to make a battery the cheapest fast-responding asset on a grid that had always paid a premium for speed. By the middle of the 2020s, at the volumes the factories were running, it had dropped under a hundred and fifty. The curve fell in steps. Each step was a manufacturing process that had been pushed past a threshold and then held there, the cell chemistry simplified to iron and phosphate so it no longer needed scarce cobalt, the cells packed into modules and the modules into cabinets that arrived on a truck ready to bolt down. By the end of the run the cells themselves had become so cheap that the largest single cost of a finished installation was the labor to put it in the ground, which is the signature, in any technology, of a thing that has finished becoming expensive and started becoming infrastructure.

Curves like this are made by manufacturing volume. For more than a century, across products as different as airframes and transistors and solar panels, engineers have watched the same rule hold. Every time the cumulative number of units ever built doubles, the cost of building the next one falls by a roughly fixed percentage, because each doubling forces a thousand small improvements in process and materials and scale. Batteries were riding one of the steepest such curves anyone had measured, pulled along by the enormous volumes first of consumer electronics and then of electric cars. Every laptop and every phone and every car sold was, without anyone intending it, an investment in making the next grid battery cheaper. The consumer products paid for the climb down the curve, and the grid inherited the price at the bottom.

The chemistry shifted underneath the curve as well. The early lithium-ion cells leaned on cobalt, a metal mined in a few troubled places and priced like the scarce thing it was, and for a battery meant to move a car the cobalt bought energy density, the most range in the least weight. A battery that sits still on the ground has no such need. Weight on a paddock costs nothing. So the stationary-storage cells moved to a chemistry built on iron and phosphate, cheaper and more abundant and longer-lived and far harder to set on fire, accepting more bulk for the energy stored in exchange for a lower price and a safer cabinet. The trade that would have been wrong for a car was exactly right for a grid, and the largest batteries in the world stopped needing the scarcest metal in them.

Running alongside the battery curve, and feeding it, was a steeper one. The price of a solar panel had collapsed even faster than the price of storage, falling by something like ninety percent over the decade as the manufacturing moved to enormous scale, until in the sunniest places electricity from a new solar farm had become the cheapest power any society had ever generated, cheaper than coal, cheaper than gas, cheaper than anything that burned. The trouble with that cheap power was only ever its timing. It arrived in a flood at midday and vanished at dusk, worth little at the moment it was most abundant. Cheap solar made the energy and cheap storage moved it through time to the hours it was wanted, and the two falling curves, crossing the thresholds of usefulness within a few years of each other, turned out to be the halves of a single machine.

There was a cost to moving energy through time, and the engineers counted it carefully. A battery gives back a little less than it takes in, losing some percent of every kilowatt-hour to heat on the way through, and it wears a little with each cycle of charge and discharge until, after enough years, it holds noticeably less than when it was new. What made the economics work was that the iron-phosphate cells could be cycled thousands of times before they faded, day after day for the better part of two decades, spreading their upfront cost across so many trips through time that the loss on each one came to almost nothing. A battery that could charge and discharge every day for twenty years was a different financial object from one that wore out in five, and the chemistry that lasted turned storage from a gadget into a piece of permanent infrastructure.

A grid had always been a real-time machine, generating each instant exactly what was being consumed that instant, because there had never been anywhere to put a surplus. Cheap storage broke that constraint. Power generated at noon could be spent at dusk. A wind farm that overproduced at three in the morning no longer had to be paid to switch off. The cost of storing a unit of energy had fallen below the value of moving it a few hours through time, and once that line was crossed the entire way a grid got built and run began to rearrange itself around the new fact.

The batteries themselves changed jobs as they got cheaper. The first grid batteries, the Hornsdale generation, were power batteries, built to deliver a large jolt for a short time, ideal for catching a frequency fault and worth little for anything else. As the cost per kilowatt-hour kept falling, it became affordable to build energy batteries, cabinets that held four hours of output or six or eight, enough to take a block of cheap midday solar and pour it back across the evening peak. The short-duration battery had proven the technology could earn its place. The long-duration battery was the one that let a grid actually run on sunlight after dark, and the only thing that had ever stood between the two was the price of a kilowatt-hour, which kept dropping.

The money in this came from a simple gap. On a grid heavy with solar, power at midday could fall to nearly worthless and power at dusk turn expensive, and a battery that bought low in the glut and sold high in the scarcity earned the difference every single day. The trade was reliable enough to pull in builders by the gigawatt-hour, each new battery farm shaving a little off the midday surplus and filling a little of the evening hole. The more storage a grid added, the smaller that daily gap became, because each new battery narrowed the very spread that had paid for it, and a market that had once swung wildly between glut and scarcity began to settle. The reward for storing energy was, in the end, the very volatility it removed, and the system grew calmer the more of it was built.

JB STRAUBEL AND THE EARLY BATTERY THESIS

JB Straubel had reached cars by way of electricity. He recalled watching power lines from a car as a child, fascinated by how electricity could be stored, moved and transformed. As a college student he wired a Porsche 944 to run on batteries in a rented garage; it had about twenty miles of range. The motor moved the car, but the short journey made the cells the part he needed to change. By 2003 he was a young engineer in California, moving among electric-vehicle hobbyists and Stanford alumni, still trying to turn stored electricity into useful motion. Straubel recalled first meeting Musk after a talk on early SpaceX at Stanford. He approached Musk with an electric-airplane idea, but their conversation turned to advances in lithium-ion batteries and the car those cells might make possible. Straubel had been trying to work out what single thing might have the largest effect on the world; energy storage together with zero-emission transport sat at the top of his list. The battery had held his attention long before a car company gave him a way to work on it at scale.

The insight that made the early company possible was to refuse to wait for a better cell. The batteries Straubel wanted did not have to be invented, because they were already being stamped out by the hundreds of millions for laptops, small cylindrical cells a little larger than a finger, cheap and improving every year on someone else's research budget. The cell was the easy part. The hard problem was governing thousands of them at once, wiring them into a pack that could be charged and drained hard without any one cell overheating and taking its neighbors with it, and writing the software and plumbing the cooling that kept the whole array alive for years. Straubel's work lived in that management layer, the unglamorous engineering that turned a commodity laptop part into something that could move a car and, later, steady a grid.

There was a limit waiting at the end of that approach, and it was supply. To put batteries into cars by the hundreds of thousands, and eventually onto grids by the gigawatt-hour, Tesla would need more lithium-ion cells than the world's existing factories could readily provide. Straubel started work on that capacity before the Model 3 had even been unveiled, when he could show suppliers a plan but no order book for the car. He later recalled repeated trips to Japan, often with Musk, to convince Panasonic to build alongside Tesla. He led the team that chose the Nevada site, built the factory and its automation, and brought it into production. He remembered the project as an absorbing problem despite the unglamorous work of finding a site, winning a supplier's commitment and making a line run. When the

Model 3 reservations arrived, the planned plant had to grow faster. The supply decisions had begun before buyers could place a single order.

What the two of them were circling, underneath the talk of cars, was the grid. A car was the thing you could sell, the consumer product that would pull battery manufacturing up its volume curve until the cells were cheap enough to do the larger job. The thesis Straubel carried into the early meetings of the company was a battery thesis wearing the costume of a car company, and the costume was load-bearing, because only a product people lined up to buy could fund the factories that would drive the cell price down to where a battery on a paddock in Australia made economic sense. The car was the lever. The grid was the weight the lever was built to move. Years into the run Straubel founded a second company, Redwood Materials, while still inside the first, to close the next loop in the same chain by recovering the metals out of dead cells and feeding them back in, because at the scale the batteries had reached, the spent ones had become an ore deposit worth mining.

The recycling was the same insight running one more turn. A lithium-ion cell does not consume its metals the way an engine consumes its fuel. The lithium and nickel and copper are all still there when the cell finally dies, in a more concentrated form than any natural ore, waiting to be pulled back out and built into a new cell. At the scale the world was now manufacturing batteries, the stream of dead packs coming back from old cars and worn-out installations had become a mine that needed no digging, a domestic source of the very metals that were otherwise fought over and shipped across oceans. Closing that loop turned the long-term supply question from a matter of what could be dug out of the ground into a matter of what had already been built.

THE MEGAPACK PRODUCTION LINE

The product that came out the far end of that twenty-year bet now rolls off a line in Lathrop, California, by the thousand. The unit is called a Megapack, and it is unglamorous on purpose, a steel cabinet about the size of a shipping container that holds nearly four megawatt-hours of storage along with the power electronics and cooling and fire suppression already wired inside, so that installing one is closer to placing a vending machine than to building a power plant. A utility orders cabinets the way it once ordered turbines, except that the cabinet shows up complete on a flatbed and is generating into the grid within weeks of the concrete curing. The factory that builds them, which Tesla calls a Megafactory, was tuned to turn out tens of gigawatt-hours of these things a year and was already being expanded past that, because the order book had stopped looking like a product line and started looking like a backlog.

The Megafactory runs the unit down a line the way an automaker runs a car. Cells arrive and are stacked into modules, the modules are loaded into the cabinet, the inverter and the cooling loop and the fire-suppression system are installed and tested in sequence, and a finished Megapack rolls off ready to be trucked to a site and switched on. The older way of adding storage or generation to a grid had been a construction project, a one-off built in place over months by electricians and engineers working from drawings. The Megapack turned that project into a product. A utility that needed storage no longer commissioned a build. It placed an order, the way it might order any other piece of finished equipment, and waited for the delivery date.

Speed came packaged with the product. A conventional power plant was a decade-long undertaking of permits and turbines and transmission. A battery farm was a field of cabinets bolted to concrete pads, and some projects could move from groundwork to grid connection within a year. A utility facing a shortfall two summers away had, for the first time, an answer it could actually build in the time it had. The lead time on new supply had always made grid planning slow, and storage was the first major source that could be ordered, delivered, and switched on inside a single planning cycle.

A single Megapack held a few megawatt-hours. A few hundred of them, lined up in rows on a fenced lot, matched the output a small power station could deliver across an evening, and a really large installation ran to thousands of cabinets spread over ground the size of a farm. The unit was modest and the multiplication was not. A grid operator no longer bought a power plant as one enormous decision. It bought capability by the cabinet and added rows as the need grew.

Mike Snyder, who runs Tesla's energy and charging business, oversaw the part of the company that almost nobody had been paying attention to while the cars drew all the coverage. The customer list for what his line produced had become a map of who runs the modern world. Utilities buying frequency response and the ability to shift solar power into the evening. Data center operators putting cabinets beside the gas turbines, holding reserve power close to a training cluster whose draw can change within a job. Mining companies powering remote sites off solar and storage that had once run on trucked-in diesel. Hospitals and military installations buying the one thing a battery gives that a grid connection cannot, which is power that does not depend on anyone else's wire staying up. A decade earlier the grid-scale battery had barely existed as a category. Now it had a catalog, a lead time, and a waiting list.

The data centers rising to feed artificial intelligence had a power problem the grid could not solve on its own, because a training cluster draws power in fierce, irregular surges, swinging from near idle to full draw and back in seconds as a computation starts and stops. A swing like that, repeated across a gigawatt of load, is violent enough to destabilize the local grid it hangs from. A bank of batteries sitting between the cluster and the grid absorbs the shock, charging when the machines go quiet and discharging when they surge, presenting a smooth and steady face to the utility while the chips behave as wildly as they like behind it. The same product that steadied a grid full of solar was now steadying a grid full of computers, and the two demands were arriving together.

That thinking ran down to the size of a house. Alongside the Megapack the company built the Powerwall, a flat cabinet that hung on a garage wall and held enough energy to carry a home through an evening or a blackout, drawing from rooftop solar by day and from the cheap night hours of the grid. It was the cells and the chemistry and the falling price of the larger product, packaged for a different buyer. A homeowner with panels and a Powerwall could store the afternoon's sun and run the house on it after dark, buying less from the utility and riding out the outages that took down the street. The product line now reached from a box on a suburban wall to a field of cabinets feeding a regional grid, and it was the same machine at every scale.

The places with no grid at all were the clearest cases. A remote mine or a small island had always made its own power the expensive way, burning diesel hauled in by truck or barge, paying heavily for each delivered unit of electricity and breathing the fumes. Solar panels and a bank of batteries changed that completely. The sun already fell on the site for free, and storage let it run the operation through the night, so the diesel generators dropped from being the whole power supply to being a backup that rarely switched on. An island that had sent a large share of its money offshore every year just to keep its lights on could keep that money at home, and the payback on the hardware ran to a few years against a lifetime of fuel bills.

THE GRID AS CUSTOMER

The manufacturing numbers had begun to change the scale of the question. In three months Tesla deployed 13.5 gigawatt-hours of storage, enough capacity to make the Megapack line more than an experiment at the edge of its car business. The physical work remained cabinet by cabinet: cells accepted and tested, cooling and power electronics enclosed, a finished unit trucked to a site where crews connected it to the local system. Repeat those steps often enough and a grid operator can plan storage as a buildable part of the network, with a delivery schedule and a crew that knows what the next cabinet requires.

The same play was running across more and more of the map. The deals had grown from megawatt-hours to gigawatt-hours, whole regional grids contracting for storage years ahead. There were buildouts across Texas and California, across South Australia where the first cabinets had proven the case, across the United Kingdom and Saudi Arabia and anywhere else that had decided to run a grid heavy with solar and wind and needed something to hold the energy still between when it was made and when it was wanted. Any government or utility that wanted a modern grid had arrived, by its own route and its own reasoning, at the same purchase order.

California showed the pattern in miniature. The state had built so much solar that on a mild afternoon it generated more than it could use and watched the wholesale price of power fall to nothing or below, and then, as the sun dropped and the air conditioners stayed on, it scrambled every evening to cover the gap. Batteries closed it. Within a few years the state had installed enough storage that on many evenings one of the major sources feeding its grid, in the crucial hours just after sunset, was power that had been made by the sun at midday and held in cabinets until it was needed. A grid that engineers had once called impossible to run on sunlight was running on that afternoon's sunlight, a few hours old, poured back out of steel boxes after dark.

Underneath the regional buildouts ran a slower change in what a grid depended on. A country running on coal and gas had to buy fuel, mine it or import it, store it and ship it and hedge against its price, and a grid built that way stayed tied to the places the fuel came from. A grid built on sun and wind and storage burned nothing once it was standing. Its fuel was free and fell on the country's own territory, and the cost had moved almost entirely into the hardware, paid once and spread across decades. For nations that imported their energy, this was a change in their standing in the world, the slow loosening of a dependence that had shaped their foreign policy for a century. The cabinets on the paddock were a small early piece of that shift, and the governments signing the largest storage contracts were the ones that had already worked out where it led.

The manufacturing to feed all of this had become a global industry weighted heavily toward China, where much of the world's cell and solar production was built. American and European builders were adding lines of their own. A cabinet assembled near its destination still depended on cells, inverters, cooling parts and trained people arriving in the right order. More places able to make and service those parts would give more grids a practical route from afternoon surplus to evening supply. The important count was not only cabinets sold, but cabinets that could be installed, maintained and trusted when the light fell.

THE FARM AT SUNSET

West Texas, at dusk. Rows of steel cabinets stand inside a chain-link perimeter, transmission lines running off to a substation a kilometer away across the scrub. The evening load rises behind the wires, and the cabinets answer it, drawing down the sun they took in through the afternoon and sending it back across the grid in a steady release. The farm is working at the hour it was built for, steel boxes giving the day back after the light has gone.

6

The Factory as Technology

THE PRESS AT GIGA TEXAS

THE MACHINE FILLS the end of the building the way a docked ship fills a slip, and when it runs it runs on a cycle of roughly eighty to ninety seconds that almost nothing on the floor is allowed to interrupt. It is an Idra die-casting press, and the number that matters about it is the force it can hold shut, which is on the order of six thousand tonnes, enough to keep a single steel die clamped against the pressure of molten metal driven into it at the speed of a thrown punch. The aluminium arrives as ingots, is melted in a holding furnace off to one side, and waits there as a silver liquid at a temperature that would vaporize skin.

A shot begins when a measured charge of that liquid is forced into the closed die through a sleeve. The fill takes a fraction of a second. The metal has to reach every corner of a mold the size of a sofa before it begins to freeze, which is why the pressure is so violent and the timing so unforgiving. Then the press holds. The die is laced with channels of circulating coolant, and for most of the eighty seconds the casting is simply giving up its heat into that steel, hardening from a liquid into a shape while the machine stands clamped and quiet.

When the part is solid the die opens along its parting line, the two halves drawing back from each other, and an array of ejector pins pushes the casting clear. A robot arm swings in, closes on the still-warm part, lifts it out, and sets it on a conveyor that carries it away to be trimmed and cooled and inspected. What comes out is one piece of aluminium about the size of the rear third of a car's floor, ribbed and contoured, with mounting points and suspension towers already formed into it. The arm clears the die, the halves close, the next charge fires. Eighty seconds, and another one.

An operator stands at a console watching the cycle, reading the temperature, pressure and timing for signs that the next part may differ from the last. The part the robot lifts out is the rear underbody of a Tesla. In a conventional version that structure would be assembled from many stamped pieces, cut and pressed and then joined with welds, rivets and adhesive along a line of fixtures and robots. The press brings much of that joining work into one shot. The operator's job shifts toward keeping the metal, die and cooling in the narrow range that lets the next large casting come out sound.

The press is the part of the operation a visitor notices, but the object that actually decides what the press can make is the die clamped inside it. A casting die for a part this size is a block of tool steel that can weigh as much as the car it helps build, cut over months on five-axis mills to a tolerance measured in the width of a hair, then drilled and laced through with cooling channels so that every region of the part gives up its heat at the right rate and the metal freezes flat and does not warp as it cools. A die at this scale can cost millions of dollars and take months to make, which is why no company casts a structural part this large on a whim. Once a company has cut the die, the shape of the part is fixed in hardened steel, and changing it means cutting the die again. The molten aluminium the die shapes is kept ready in a holding furnace beside the press, a silver pool, dosed out one measured charge at a time, so the metal never waits long enough to pick up the gas and oxide that would leave the finished part full of hidden flaws.

What the robot arm sets on the conveyor is a rough draft of an underbody. The casting still wears the gates and runners where the metal flowed in, frozen stubs of aluminium that have to be cut away, and it has to be checked for the flaw that haunts every large casting, which is porosity, the small voids left where trapped gas or shrinking metal opened a bubble deep inside the part where no eye can see it. A weld on an assembly line can be inspected from the outside, its bead read like a signature. A void at the heart of a casting the size of a car floor can be found only by sending the part through an X-ray or a CT scanner and reading the shadows. So the line that follows the press is partly a foundry's trimming bay and partly a radiology department, and a casting that fails the scan goes back to be melted down and poured again, its expensive aluminium losing nothing but the energy it took to shape it the first time.

The shot lands with a sound felt in the chest more than heard, the heat off the holding furnace reaches across the aisle, and the floor carries a faint tremor on the eighty-second beat that a body learns to expect the way it learns a clock. The one person assigned to the machine is there for the moment it does something it has not done before, a temperature drifting, a cycle running long, a casting coming out a hair out of tolerance, the small early signs that a die is wearing or a furnace is off. The press does the work. The human watches for the future.

A press that can hold six thousand tonnes shut is not a thing a company buys off a shelf, because until recently almost no one had asked for one. Die-casting machines had topped out far smaller, sized for engine blocks and gearbox housings and the other dense little parts that were all anyone had thought to cast. Tesla went to Idra, an Italian press-maker, and in effect commissioned a much larger class of casting machine, a press tall as a house that could fill and freeze a mold the size of a car's floor on a clock measured in seconds. Once the machines proved out, the press itself became a product other carmakers lined up to order, and a corner of the machine-tool industry reorganized around demand that Tesla had brought into being.

At an eighty-second cycle, a press can make many castings over a shift, provided the die, metal supply and inspections keep pace. Each sound casting removes a series of separate stamped pieces and joins from the route to a finished underbody. The work does not vanish into the machine. It moves into alloy control, die preparation, trimming, scanning, maintenance and the decisions an operator makes when the trace drifts. A shorter line can give the building room for another operation, while a bad casting sends its material back to the furnace and its measurements back to the people who can change the process.

THE SINGLE-PIECE MOVE

The press was the visible end of a decision that had been made years earlier and several layers deeper, and the decision was about sequence. The way a part gets made in the automotive and aerospace industries had been settled for a century. An engineer designs the part to do its job, freezes the design, and hands it to the manufacturing people, whose problem is to figure out how to build the thing they have been given. Tesla designed the vehicle around the casting, not the casting around the vehicle. The shape of the part, its alloy, the rate at which it could shed heat without warping, the draft angles that would let it release from the die cleanly, all of it had to be solved together with the press that would cast it, as one coupled problem with one set of answers.

The hardest piece of that coupled problem was the metal itself. The aluminium alloys the casting industry had used for decades reached their full strength only after a part was baked in a high-temperature heat treatment, and a part the size of a car's floor, baked that way, would slump and warp as it cooled until it no longer fit anything. Tesla's materials group spent much of the development running down a different answer, an alloy that came out of the die already strong enough to carry the loads a car puts through its underbody, without a later heat treatment that could warp a casting of this size. The alloy was invisible in the finished car and was as much the enabling invention as the press, because without it the single-piece floor was a part that could be cast and could not be used.

A car's underbody is not only structure, it is a crash structure, engineered to fold in a controlled way and soak up the energy of an impact before it reaches the people inside. The old welded assembly let engineers tune that behaviour part by part, making one panel thicker, one rail softer, choosing exactly where the structure would yield. A single casting gave up that fine control, and cast aluminium carried a reputation for being brittle, for cracking under a hard blow where a folded sheet-metal rail would crumple and absorb it. Getting a casting to crumple the way a crash needs it to, through the geometry of its internal ribs and the ductility designed into the alloy, was a harder-won achievement than the press.

It took a mountain of scrap to get there. The first big castings cracked as they cooled, warped out of tolerance, came off the line riddled with the porosity the scanners were built to catch, and for a long stretch the reject pile was the most reliable thing the press produced. The team revised the die and the alloy and the cooling together, because they were one problem, changing the geometry of a rib here and the chemistry there and the timing of the shot, casting and scanning and melting the failures back down, until the part came out sound often enough to build cars from. The press on the floor at Giga Texas was the survivor of that grind, a machine that ran because a few hundred castings before it had not.

The same drive to collapse parts into one reached the battery as well. Tesla made the pack itself a structural member of the car, its case forming part of the floor, so the seats bolted directly to the top of it and the two big castings, front and rear, joined to its ends. A car built this way did away with the separate floor pan laid over a separate pack. Three large objects, two castings and a structural pack, bonded into one rigid body, and the count of pieces in the most important structure of the car fell from the hundreds toward the low single digits.

The single casting carried a cost on the other side of the ledger, and Tesla accepted it with open eyes. A welded assembly of small parts can be repaired in pieces, a bent rail cut out and a fresh one welded in after a collision. A casting is one object. Damage it badly enough and the repair is to replace the whole part, or to write off the car, which pushed up the cost of some crashes and raised questions about repair costs after a wreck. Tesla judged the manufacturing gain worth the repair penalty and kept casting larger.

The single casting also reached back up the supply chain and pulled it in. The roughly seventy stamped parts it replaced had required tooling, handling, joining and transport, and between every one of those vendors and the line sat buffer inventory, racks of pressed panels waiting to be welded, a web of contracts and shipments and warehouses that existed only because the underbody had been a kit. One shot of aluminium collapsed that web. The joined assembly now emerged as a single casting on Tesla's own floor, and a long stretch of the company's supply chain simply ceased to be something anyone had to manage.

During the worst of the Model 3 ramp, one throughput problem ran through a two-million-dollar robot cell that set fiberglass mats between the battery modules and the floor pan. The team made the robot faster. They gave it more torque. The cell still could not keep up. So Musk asked what the mats were for. The battery-safety group said they damped noise and vibration. The noise-and-vibration group said they were there for fire protection. Neither group had installed them. Neither could name who had. The requirement had outlived whoever set it, and it was being kept alive by two teams each pointing at the other. They put one car with mats and one car without on a test, microphones in both, and could not hear the difference. The mats came out, and the robot cell came out with them. The order he later named for this work ran question the requirement, delete the part, simplify what is left, speed it up, and only then automate it. On the mat line the company had run that order backward. It had automated a step first and asked whether the part should exist at all only last.

Knowing how much of all this to hand to a machine was its own hard lesson, and Tesla had learned it the expensive way. In 2018, at the battery factory in Nevada, Musk had tried to automate the production of battery packs almost completely. Robots were to grip the cells, robots were to place the modules, robots were to run the conveyors that moved the work from station to station. The line seized. The robots were slower and less reliable at the fiddly work than the humans they had replaced, and the whole plant fell behind. Late at night, Musk and a small group around him walked the line and started tearing the automation out by hand, cutting machines that were not earning their place out of the conveyors and hauling them off the floor. Omead Afshar and Antonio Gracias were among the people walking the floor in those weeks. The machines that failed the throughput test were removed. That spring he said the lesson in public, in his own words. Excessive automation at Tesla was a mistake, he wrote, his mistake to be precise, and humans were underrated. The principle that came out of that year was short enough to paint on a wall. Put people on the line first, learn what the work actually is, and hand a motion to a machine only once it has been understood. Eight years later, on the floor at Giga Texas, the press was what that principle looked like once it had hardened into steel.

By 2018 Musk had made the production floor a center of his attention. He spent his nights on the production floor, he put some of his strongest engineers on the question of how to build the car, and he reviewed the line with the intensity that the rest of the industry reserved for the product. By the standards of the Detroit he had walked into, this was close to heresy, because the industry had always treated the product as the achievement and the plant as the place you bolted it together.

SANDY MUNRO TEARS DOWN A MODEL Y

Sandy Munro had spent his career taking cars apart. His firm bought finished vehicles on the open market and stripped them for automakers who wanted to know how each assembly was made and what it cost. He knew Detroit manufacturing from the inside; his method began with the part on the bench.

A teardown was methodical work. A car went onto a bench, and over weeks Munro's team took it down to its last bracket and bolt. They weighed parts, measured joints, traced how each assembly had been made, and estimated the labor needed to put it together. The reports turned a finished vehicle back into parts, tooling and line steps that another builder could examine rather than merely admire.

His early reviews of Tesla had been hard on panels that did not line up, doors that shut wrong, and paint and trim that looked rushed. When he later took apart the Model Y underbody, he could examine the casting as a manufacturing choice on its own terms. The panel fit still mattered to a buyer. The casting showed what a different process could do to the number of parts a plant had to hold, locate and join.

He took a Model Y apart and found a casting where many stamped pieces and joins would otherwise have been. On his channel he could point to the ribs, mounting points and continuous metal, then count the fixtures and welds that were no longer needed downstream.

The part count alone did not make the new method transferable. A buyer could order a large press and study the Model Y's ribs; neither act recreated the qualified process on the original line. The earlier trials could guide another plant's work, but that plant still had to qualify its own castings. What traveled was a method for making linked design and production decisions, not a secret shape to copy.

Other automakers began testing large castings and the processes around them. A press could be purchased; learning where a rib should thicken, how a die should cool and when a void made the part unusable took its own trials. Each plant had to decide which assemblies were worth collapsing and which still benefited from smaller replaceable pieces. The method could spread without every builder making the same car.

THE "UNBOXED" PROCESS

In 2023 Tesla showed where the casting was heading. The company called it the unboxed process, and the name pointed at what it abandoned, which was the box, the painted body shell that every car since the 1910s had been built inside of and around. The moving assembly line that Henry Ford had set running at Highland Park in 1913 carried a car-to-be down a single path, and every part got added to it in sequence, one station after another, the whole car held hostage to the slowest step on the line. Tesla's proposal was to stop building the car in a line at all. Six major modules, the front, the rear, the two sides, the floor, and the fully fitted interior, would be built at the same time in parallel bays, each one open and easy to reach because it was not yet boxed inside a body. Only at the very end would the modules be brought together and joined into a finished vehicle at a single combining station.

A worker or a robot fitting out the interior of a conventional car has always had to reach in through a door opening to do it, twisting to set a part deep inside a shell that is already painted and sealed and awkward to work within. An unboxed module sat open on every side, easy to reach, easy to hold in a fixture, easy to hand to a machine that could come at the work from any angle. Painting accessible modules before the final join could loosen the single-file bottleneck of a conventional paint shop.

Tesla unveiled the unboxed idea at its 2023 investor day and tied it to a specific ambition, a next-generation vehicle cheap enough to be bought by most of the world. The way to reach that price ran through the line. The company set itself the goal of cutting the cost of building a car by roughly half, and the unboxed process, with its parallel bays and its proposed parallel paint step and its fewer, larger pieces, was the machine meant to deliver that number.

Lars Moravy, who ran vehicle engineering at Tesla, described the unboxed process as a way to alter how people and tools reach the car while it is being made. A module open on several sides can be worked on before it meets the other modules. The team would still have to learn where the bays wait on one another, how to move painted parts without damage and how to check the final joins. A drawing can set out the intended sequence; only a running line can reveal which operation holds the others up.

Tesla proposed that parallel modules could reduce the floor space and cost of a plant while increasing output. The difficult test would come where the bays meet. If one module arrived late or failed inspection, the joining station would wait even while the other bays kept working. A line built this way would need visible buffers, sound handoffs and workers able to change a bay without losing the rest of the day's flow.

THE FACTORY IS THE PRODUCT

The car remained the thing a driver needed, but the method for making it had become a design problem in its own right.

A casting changed the number of parts reaching assembly. A reachable, open module might change the time and strain of installing them. A plant could learn from both only if the people on the line were able to find what delayed or damaged the next car and change the work that caused it.

This too had been learned at the limit, in a single desperate month. In the spring of 2018 the Model 3 line at the Fremont plant had stalled, and the company was burning cash it did not have while it failed to build cars fast enough to survive. Analysts were writing that it would run out of money before it could build its way to safety, and on the numbers they were close to right. The automated line in the main building, the one meant to prove that robots could assemble a car start to finish, was choking on its own complexity and producing a trickle where the plan had called for a flood. Musk and Jerome Guillen put up a tent. It was a thousand feet long and a hundred and fifty feet wide, raised in the parking lot under a permit for a temporary vehicle repair facility because a permit for a real assembly line would have taken months Tesla did not have. Inside it they ran a general assembly line that had almost no automation at all, a repurposed parts conveyor tilted down a slope to feed the work along by gravity, with people standing at every station doing by hand what the robots in the main building could not be made to do reliably. The first car came off the tent line on the sixteenth of June. By the end of that month Musk had taken to sleeping on the factory floor, and one night near two in the morning he watched the five-thousandth Model

3 of the week roll out. He wrote afterward that he thought they had just become a real car company.

The Nevada and Fremont lines taught a practical sequence. People had to make the operation run, see where it stopped, remove steps that served no purpose and give a machine the motions it could repeat reliably. Under the tent, workers carried the changing work while the line was still being understood. Later machines could take the settled motions, leaving the crew to find the next constraint. The lesson did not make a finished factory permanent; it made revision part of operating one.

The layout was part of the technology. A teardown could show the dimensions of a casting; it could not show why one station waited, why a worker reached around a fixture or which measurement made the next shot safe to pour. Those answers lived in the line and in the people changing it. The tent had shown that a crude arrangement with clear work and human hands could make cars while a more elaborate line stalled. The die-casting line showed what could be handed back to machinery once the work was understood. Both methods could be learned by other builders willing to read their own floors as closely.

THE LINE THAT KEEPS CHANGING

At the end of a shift, the line carries the day's changes into another crew's hands. A fixture has moved, a check has been added, or a part once carried around a stopped machine now passes through it. The next workers need to know which change held and which still needs watching. A factory able to make that pass-down repeatedly can make more than one design well; it can keep learning how to make the next one.

7

The Labor Ceiling

THE ROBOT ON THE LINE

GIVE A TESLA OPTIMUS the eighteen-meter walk between the pack line and the integration station at Fremont. At nine in the evening the floor is under the same white light as at nine in the morning. Its gait is bipedal and deliberate, the weight settling onto each foot with a half-beat of hesitation before the next step. That walk has to be trained into the machine before it can be trusted with a load.

Its job on this line is simple and exact. A finished battery pack comes off the pack line on a powered cart, and the machine takes the cart to the integration station, the weight riding on the wheels while the machine steers and steadies the load, then guides the pack into place and walks the empty cart back for the next one. It makes the trip again and again. A few stations down, a worker makes a version of the same walk with parts the machine cannot yet handle, sees when a part sits badly on the cart, and puts it right before the next handoff. At shift change the worker goes home, another crew comes on, and the robot keeps walking its eighteen meters. The people on both shifts attend to the parts that do not yet fit the motion it has learned.

There is a short handoff sequence built into the route. The pack line sets a finished pack onto the cart at a known height and in a known position, and the machine learns that handoff along with the floor. It steps in, squares itself, closes both hands on the cart at the points made to be gripped, and walks the load to the station, the cart bearing the weight while the machine holds it true. The cart, grip points and clear walking lane are part of the work. They can be laid out once and used by another unit trained for the same route.

The machine makes very little noise. There is the soft servo whine of the actuators as the legs cycle and a faint sigh from the torso when it bends to set the pack down, and under the general noise of the line you have to be close to hear either. It does not announce itself. A visitor walking the floor for the first time tends to catch sight of it late, registering a person at the edge of vision and then understanding, a half-second behind, that the thing carrying the pack has no face turned toward the work and no weight shifting from boredom and no eyes that drift toward the clock.

The worker a few stations down has the night shift, and at a break the crew talks about the machine as they would any new equipment, half watchful and half indifferent. It still goes down. A gripper fault or a low charge pulls a technician onto the floor to stand it back up, and the line absorbs the gap as it would any stoppage. The machine is not yet better than the people around it at most of what they do. It is partway through its training, and the people responsible for the line can see exactly where that training runs short.

A bipedal machine doing a manual job alongside people had been, within living memory, a thing that lived in films. Here it has a route, a gripper fault and a technician who comes when it stops. A serial number, a maintenance log and a handoff the next shift knows would make it part of the day's work. The remarkable step is a motion ordinary enough for a crew to rely on without having to watch it as a demonstration.

THE HAND

What lets it attempt the job is at the end of its arm. The Optimus hand has articulated fingers and a thumb, with tactile sensing designed to report how hard a fingertip is pressing and where. In an early demonstration, the hand closed on a battery cell and seated it in a tray. A wrench, a small fastener and a cell each ask for a different grip. The fingers have to change pressure as the surface gives or slips, because the same force that holds a wrench could score the soft casing of a cell.

For most of the history of industrial robotics, the reliable factory grip was a grip designed for a known part in a known place. An arm could lift a car body with a fixture made for that body; a small fastener arriving at a new angle demanded a different tool and setup. Two steel fingers or a vacuum cup could repeat a motion with great force and precision, while mixed, soft or misplaced objects made the task harder. Closing the gap between strength and delicate adjustment is one of the engineering problems a humanoid hand must solve. The hand has to fit on a wrist, work through many kinds of part and be made again for the next machine.

The difficulty was familiar in the field. The tasks that came easiest to a person, closing a hand around an unfamiliar object and settling the grip by feel, had turned out to be the hardest tasks to put into a machine, and the tasks a person found hard, arithmetic and logic and the recall of long lists, had turned out to be the easy ones. A computer had been beating grandmasters at chess for thirty years before a robot hand could reliably handle many ordinary objects. The hard part lives in the loop between touch and force, the constant adjustment a human hand makes as it settles onto a thing, easing off where the surface gives and bearing down where it holds firm. The Optimus hand is being built to close that loop in silicon. A fingertip sensor would report pressure; a controller would read it against the grip the task needs; the fingers would have to correct before the cell deforms or slips, inside the motion of setting the part down.

The hands that worked in university labs and corporate research wings were often one-offs. A technician could machine, wire and calibrate each one by hand because a research instrument only had to exist in small numbers. A machine meant for thousands of floors needs a different kind of hand. The actuator count, tendon routing, tactile sensors and assembly time all have to become repeatable work for a production line, while the finished hand still feels a cell gently enough to place it. Tesla is trying to make that second achievement. The line must produce a calibrated hand again and again, with the same useful touch in the first unit and the hundredth.

The choice to make the machine human-shaped at all was a choice about the world it would work in. A factory floor, a warehouse aisle, a set of stairs, a doorway, a workbench at standing height, a tool sized for a human grip, all of it had been built across two centuries to fit the human body. A wheeled robot needs ramps where there are stairs. A fixed arm needs the work carried to it. A machine that walks on two legs and reaches with two arms can attempt work in more of those spaces, though the route, safety boundaries and handoffs still need to be made sound. The price of the human shape is bipedal balance. Walking upright across an uneven floor while carrying a shifting load is a continuous act of catching a fall that never quite arrives, and the fast loop of sensing and correction has to run through the whole body as well as the hand.

To reach a price at which many machines could be built, Tesla designed much of the body itself. The actuators that drive the joints, small dense motors and gearboxes standing in for muscle, can reuse designs across several joints in the body. An off-the-shelf robotics actuator is a low-volume specialty item priced like one. Reusing a limited set of designs across each body gives a supplier a longer run and gives the assembly line fewer distinct parts to hold, inspect and replace. The body is being engineered backward from a target price, much as the casting that formed a car's underbody was engineered backward from a faster line.

The hand is good in demonstrations. It is still learning the range of grips a human hand makes without thought. Slick surfaces, irregular shapes and coordinated two-handed work demand changes in sensing, control and practice, and the engineers know that list because closing it is their work. A grip that passes from a demonstration into a repeatable station would take one repeated motion off a person's shift. The awkward part, the failed handoff and the next change in the work would still ask for someone who can see what the machine missed.

THE OLD INDUSTRIAL ROBOTS

The use of that hand comes into focus alongside the machines the world's factories already hold. About five million industrial robots were already operating worldwide. They are made, in large numbers, by ABB out of Switzerland and Sweden, Fanuc and Yaskawa out of Japan, and KUKA out of Germany, the last of them owned since 2016 by a Chinese appliance maker. A familiar form is the six-axis arm, bolted to a pad and assigned a fixed station. It welds the same seam, places the same component or sprays the same panel through long runs, after an engineer and an integrator have taught the path and built the cell around it.

There is a reason many fast welding arms live behind fencing. An arm moving at full speed through its programmed arc carries enough force to injure a person who steps into it. The conventional cell keeps the person out of that envelope with steel mesh, light curtains and interlocks that stop motion when the space is entered. Newer systems can sense people or work nearer to them, but the guard remains a practical part of many high-speed installations. Parts have to be staged through the boundary, and the space given to the cell has to be counted along with the arm itself.

Those arms cluster in automotive welding and electronics assembly, where the same part returns in the same orientation through a long run. A shop making varied parts has to interrupt that work to change the fixture and program for the next run. Other industrial machines already move through warehouses and some arms can be reprogrammed, so the boundary is practical rather than absolute.

The price of the arm was only part of the price of automation. A systems integrator had to build the fixtures, guarding and safety logic around it, then check the path it would follow. That work took time and money before the arm made its first sound part. Over a long run it bought speed and precision; on a short one, the setup could consume the time the arm was meant to save.

WHAT GENERAL-PURPOSE UNLOCKS

The jobs that got left to people were rarely the tasks that drew attention. They were the second fastener that sat at an angle the welding cell could not reach, the tray of mixed parts that arrived in no fixed orientation, the once-a-week changeover that came too seldom to earn its own machine, the bin of returns that each needed a slightly different touch. A dedicated machine could have been built for any one of them, and building it seldom earned its keep because the task did not repeat often enough or uniformly enough to pay back the cell. A person brought the adjustment that cell lacked, and made it again for the next part and the next.

A trainable hand and a body that can move between stations would give a factory a different tool for that mixed work. The same machine might carry a cart on one shift, present a tray to an operator on another, and tend packaging when the run changes. Each assignment would need demonstration and a safety check, but it would not require building a new fenced cell around every small job. Between the dedicated machines, people now carry, sort, replenish and correct the flow as parts change. A mobile hand could take some of the repeated load; the crew would judge it by what happens when the next part arrives at the wrong angle.

The machine itself would have to be manufactured, maintained and taught at volume. A hundred bodies coming off a line would be a hundred possible workstations, not a hundred finished abilities. Each would need charged batteries, sound joints, a route, safe handoffs and tasks whose failure a technician could find and repair. That is still a large change from a one-off research hand. Once the same capable body can be built repeatedly, a small factory or warehouse can plan to add handling capacity without designing a new special-purpose cell for every task. The number of useful machines would grow only as fast as the tested motions, service crews and places ready to use them.

The second job begins with teaching. The fixed arm's path is specified and checked against a fixture built for one task. A more adaptable machine can learn from a person demonstrating a motion or guiding its hands through it, then run trials against parts placed at slightly different angles. The engineers examine where it fumbles, alter the policy, and test again. Software carries much of that lesson, while a new station may still need a different cart, grip point or safety boundary. The robot that moved packs in a trial could be taught a packaging task later, but only after the new motion held through enough real cycles. The body is reusable; the work of making its next skill dependable remains.

Price decides how far beyond a laboratory or a large plant the hand can travel. Tesla is aiming for a price below thirty thousand dollars for an Optimus built at volume. The purchase price would be only part of a working station. Power, maintenance, the technician called after a gripper fault, safe integration and the time spent teaching each task all belong in the sum. Against those costs sits the possibility of using one body for several short or irregular jobs that would each be too small to justify a dedicated cell. If production makes the hand repeatable and the trials make its skills dependable, a smaller operation could use the same kind of mechanical help that only a heavily automated plant could once afford.

Tesla plans to send its first Optimus builds to an Academy for training. Its own factory is a place to watch a robot closely, with engineers able to study a failed grip, a blocked route or a task that takes too long for the line's takt. The failure can be carried back to a hand design, a policy or a change in the station. A motion that holds through repeated cycles can enter the schedule, while the tasks the machine cannot yet do remain in front of the engineers who are paid to close that gap.

THE WORK BEYOND ONE STATION

Underneath the casting press and the line that keeps changing, a great deal of work still happens between stations. Someone brings a cart, sorts a mixed tray, sets a part back when it arrives wrong and notices a fault before the next unit inherits it. On a night shift those acts ask for a person's body and attention through hours of repeated lifting and carrying. A trainable hand might take a defined share of that load. It would be useful only if the line could still count on the person, the technician and the machine to catch what the next pack needs before it moves on.

The boundary between a machine's repetition and a person's judgment has moved through the history of the factory. A loom took over a long sequence of thread movements, while people still repaired breaks and watched the cloth. A furnace grew larger and more controlled, while someone still had to read its condition and intervene when the material changed. A car line handed welding and spraying to fixed arms, then relied on people to bring varied parts to those cells, change the work over and respond when a join failed. Each machine altered the pattern of labor around it. A mobile hand able to work across several small jobs would move that boundary again, and the result would depend on what the hand could actually do through a long shift.

The eighteen-meter cart route is one task. A battery pack carries cells, housing, cooling, controls, assembly and inspection as well as that walk, so moving the cart alone does not set the pace of the whole line. It can clear a repeated carry between the pack line and integration station if supply and final test are ready for the next unit. The tasks beyond that route ask more of the hand because the object and the timing change. A grip taught for a mixed tray would have to work on the part that arrives out of position as well as the easy one. If it does, the line gains handling capacity at a point where a dedicated machine was never worth installing.

A narrow version of that movement is already working in a warehouse. At GXO's Flowery Branch facility, Agility Robotics' Digit takes totes from mobile robots and puts them on a conveyor under a commercial agreement. The deployment had passed a cumulative 100,000 totes. The robot handles a defined handoff, and when downstream work backs up, the system can set totes aside and load them later. A machine that recovers from that ordinary interruption has begun to help with the variation that makes mixed work difficult.

The cart route continues through a crew change. Cell supply, integration and final test set the pace at which good packs leave the building; the route adds output when those neighboring stations can accept the next unit. A hand that learns replenishment or a changeover could remove a waiting interval that stops the line. More good packs would then cross final test on the same floor and tooling, and the crew would see where the next constraint has moved.

The place a factory can stand depends on more than the cost of an arm or a wage. Power, parts, trained people, service crews and the distance to the user all enter the decision. A small plant handling varied orders cannot spread a dedicated cell over millions of identical pieces, so it often depends on manual handling between the machines it can afford. One week the order might be a pump housing, the next a bracket for a storage cabinet. The cutting, casting and inspection would still need their own tools and people; a trainable hand could keep material moving between them as the order changes. If that work becomes affordable in many small plants, more equipment can be made or repaired near the place it is needed.

The arrival of a useful humanoid still runs on several clocks. Processors have to read a changing floor quickly enough for balance and grip. Actuators and batteries have to fit inside a body that can work and be repaired. The training methods borrowed from autonomy have to meet a hand, a cart and a part that arrives at the wrong angle. Tesla is building toward that combination, and Ashok Elluswamy's team carries the AI work for Optimus. A machine that can repeat one safe route and then learn another would give a workshop or a warehouse an adaptable piece of physical equipment. The next layer of difficulty is visible in the hand itself, when an object shifts and the machine must decide what it is seeing before it closes its fingers.

PART III

THE INTELLIGENCE LAYER

8

The Unprotected Left

THE UNPROTECTED LEFT

PICTURE A WEEKDAY evening in San Francisco's Mission District, a Tesla Robotaxi at the head of a left-turn lane, its nose easing toward the centerline. At six forty-seven the intersection is doing what it always does at that hour, which is too much at once. A safety driver sits at the wheel, watching the system's proposed turn; a passenger in the back looks down at his phone. The car is waiting to cross two lanes of oncoming traffic under a flashing yellow arrow that grants permission without granting protection. On the near corner a small crowd waits to cross, one person stepping off the curb early. A cyclist is coming up the right side from behind, fast, in the car's blind quarter. Oncoming traffic comes in clumps with short gaps between them.

An unprotected left gathers several of a driver's hardest judgments into one turn. It asks a driver to judge the speed of oncoming traffic, find a gap in it, and commit the car across the path of vehicles moving at speed, all while watching for the pedestrian and the cyclist who have every right to be exactly where the car needs to go. A large share of the worst urban collisions happen in this one geometry. It is the maneuver a self-driving program is judged on, because getting it right turns on what had been hardest of all to build, a working judgment about what other people are about to do.

A human making this turn does a great deal without knowing he does it. He reads the oncoming driver's speed from the way the car grows in the windshield, guesses from a glance whether the pedestrian on the curb has seen him, keeps half an eye on the cyclist he noticed a block back, and holds all of it together while his foot rests on the brake and his hands wait on the wheel. He has learned this over years and runs it below the level of words, and on a good day he is very good at it. On a bad day he is tired, or looking at his phone, or sure the gap is wider than it is, and the unprotected left becomes one of the most dangerous ordinary things a person does in a car. The machine has to be at least as good as the human on his best day, every time, including the times the human would have been on his worst.

The car reads the intersection through cameras alone. Eight of them ring the body and look out in every direction at once, and from those flat streams of pixels the network builds its own moving model of the space around the car, the distances and the speeds and the paths, with no spinning laser on the roof to measure depth directly. The bet behind that choice was that a system made to see the world the way a person sees it, from light, would in the end be cheaper to build into every car and more general than one leaning on expensive specialized sensors. It made the problem harder in the near term and the hardware cheaper for good.

The car is tracking all of it. Across its eight cameras it must keep the oncoming cars, the early-stepping pedestrian, the cyclist in the blind quarter and the delivery truck double-parked twenty meters up the cross street in view. It needs their positions, speeds and likely paths as the scene changes; the truck will narrow the lane the car is turning into. The streams keep changing, and the car must turn them into one decision in time to move. When is the gap.

Behind the stillness of the waiting car, several outcomes remain possible, and the scene can change before any of them has played out. There is a version where the pedestrian commits to the crossing and the car holds. A version where the cyclist wobbles into the turn path and the car waits him out. A version where the gap that looks open is shortened by a driver accelerating to beat the light, and the turn is broken off halfway through. The car has to revise its decision as the scene changes, and the clean opening might last only a moment.

How much room is enough. The car needs a gap long enough to cross two lanes and clear the path of the nearest oncoming vehicle before it arrives, which at city speeds is a matter of a few seconds and a few car lengths, and the cost of getting it wrong is not symmetric. Wait for too much room and the car sits through opening after opening, the man in the back grows impatient, and a line of traffic builds behind it. Take too little and there is a collision. The network has to find the smallest gap that is safely large, and find it in the half-second before it closes, and it has to make that same call thousands of times a day across millions of cars without once getting the cheap impatience and the catastrophic haste confused.

A clump of oncoming cars clears. For a little under two seconds there is room. The cyclist has drawn level and is past. The early pedestrian has thought better of it and stepped back. The network commits, the car turns, smoothly threading the gap and settling into the cross street already accounting for the double-parked truck ahead, and the whole thing is over. In his seat the man glances up, registers that the car is now pointed a different direction than it was pointed before, and goes back to his phone. He has not thought about the turn at all, because there was nothing to think about, because the machine did the thinking and the machine was right.

FROM SOFTWARE TO WORKER

For the first decade of the modern machine-learning era, the artificial intelligence that the public met was something that produced symbols on a screen. It talked, in the language models. It drew, in the image models. It ranked and recommended and predicted the next word and the next purchase, and for all the power of it, the work it did happened inside a data center and came out as text or pictures or numbers that a human then acted on. The second decade is producing a different system. It drives. It could learn to move a battery pack between stations. It picks an item off a shelf and packs it in a box and welds a seam. The name that has settled on this change is embodied AI, and the name is pointing at a shift in where the intelligence lives. The intelligence has moved into the thing that does the work. The AI is the vehicle. The AI is the robot. The AI is the manufacturing cell, and the software has become the means by which a body acts in the physical world, with the body as the artefact the software leaves in the world when it acts.

The first decade had felt like the whole story while it was happening. Each year brought a system that did something with words or images no machine had done before, and the public argument was about what those systems meant for writers and artists and the truth of a photograph. Underneath that argument a quieter shift was loading. The same methods that had learned to predict the next word were being pointed at the next motion of a steering wheel and the next placement of a robotic hand, and the systems that learned those things were already climbing down off the screen, into the car, onto the factory floor, out into the street.

The move into the body changed what an error costs. A language model that picks a wrong word writes a clumsy sentence. A recommendation engine that guesses wrong shows someone a film they switch off after ten minutes. Those mistakes live on a screen and are undone by a human who shrugs and scrolls on. The same statistical system, with weights grown from examples, now sits inside a two-ton car at a city intersection, and there the distance between a good output and a bad one is measured in crushed metal and broken bone. Embodiment raised the cost of being wrong by as much as it raised the value of being right.

An intelligence that acts in the world also has to keep its world's clock. A language model can take a second or two to answer and no one minds. The network at that Mission intersection had no such grace. It had to take in the scene, weigh its branches, and commit the car in the half-second before the gap it was reading closed, and it had to do that continuously, the perceiving and the deciding and the acting folded into one loop that ran many times a second as long as the wheels turned. The world it worked in did not pause to wait for it to think.

What carried across all of these machines was a single way of building them. The network that decided when to take the gap and the network that told a robotic hand how to close on a battery cell were different networks of the same kind, each grown from examples of its own task, each running the same computation of weights and signals. A company that learned to grow one of them had learned most of what it needed to grow the others. The driving program and the walking program and the program that would one day run a packing line were branches off one trunk of method, which is why a carmaker found itself, without quite changing businesses, in the robot business and the AI business at once.

KARPATHY'S OLD ESSAY

There had been a warning of this, stated plainly years before most of the industry was ready to hear it. In 2017, Andrej Karpathy, then the director of AI at Tesla, wrote an essay in which he proposed that the nature of software itself was changing. The software the world had been writing for seventy years was written by people in languages people could read, instructions typed one line after another by a programmer who understood what each line was for. Karpathy called that Software 1.0. The new kind was grown. A network was shown millions of examples of a task and adjusted its own millions of internal weights, over and over, until it produced the right behavior, and the program that resulted existed as a vast matrix of numbers that no human had written and no human could read. He called that Software 2.0.

The difference showed up most sharply in how you fixed such a program when it misbehaved. A Software 1.0 program that did the wrong thing had a line of code somewhere that was wrong, and a programmer could in principle find that line, read it, and correct it. A Software 2.0 network that did the wrong thing had no such line. The wait for the cyclist and the yield to the pedestrian stepping off the curb had been written by no one. The behavior had been grown from examples of those situations, and the way you corrected it was to change the examples, to feed the network more and better instances of the thing it was getting wrong and grow it again. The engineer's craft moved from writing instructions to shaping the data those instructions were grown from. The most valuable thing a company in this business could own stopped being a codebase and became a library of the world.

This was a harder thing to trust than it first sounded. A program a person could read could be checked by reading it. An expert could study the code that flew an aircraft or ran a pacemaker, follow its logic line by line, and satisfy himself that it would behave in cases no one had yet tried. A grown network offered nothing to read. Its competence could be observed, measured, counted in miles driven and collisions avoided, but it could not be proven the old way, by inspection, because there was no chain of reasoning inside it to inspect. A machine of this kind earned trust the way a person does, by behaving well across an enormous number of situations until the watching parties were satisfied, and that change in how trust gets established weighed most precisely where the network had moved, into the bodies that could hurt someone.

The essay was treated at the time as an interesting provocation from a young researcher, something that circulates in the field and is filed away. Nine years later it was the operating doctrine of an entire industry. By 2026 the great majority of the behavior that ran a Tesla, the part that decided when the gap at that intersection was wide enough to take, had been grown from the driving of millions of cars. It lived in the weights, and the engineers who worked on it spent their days shaping the data and the training that grew it.

Grown from enough miles, the driving network learned things nobody had set out to teach it. It learned the small social grammar of traffic, the way a car noses forward an inch to claim a gap, the difference between a pedestrian who is waiting and a pedestrian who is about to step, the particular hesitation of a driver who is lost. It is the tacit knowledge a human picks up over years behind the wheel and could never fully put into words, and it had come into the network the same way it comes into a person, from seeing the situation happen many thousands of times and absorbing how it tends to go.

The pattern had repeated across every domain the new methods touched. Machine vision went this way first, when networks grown from labeled images overtook decades of hand-built rules for recognizing a face or a street sign. Speech recognition followed, then translation, then the language models the public would later meet. Each field had a stretch of years where the hand-coded approach and the grown approach ran side by side, and each stretch ended the same way, with the grown system pulling ahead once it had enough examples and enough training behind it. Driving was the same story arriving in a heavier and more dangerous medium. Judgment in traffic had to be grown, and the examples it had to be grown from were miles.

THE FLYWHEEL

What grew it was a machine for turning miles into intelligence, and the machine had a name inside the field. The flywheel. Every Tesla on the road is a sensor platform that never stops gathering, and by mid-2026 the fleet running this software numbered in the millions of vehicles, in dozens of countries, across cities, highways and changing weather. When one of those cars meets something its network handles poorly, an unusual merge, a strange piece of road furniture, a near-miss at an intersection, the event can be flagged, the relevant seconds of camera footage uploaded, and the case added to the enormous library of situations the next version of the network will be trained against. The next release drives the situation better. Then it goes back out across the fleet and gathers the next set of cases it does not yet handle well.

By 2026 the grown part of the system had eaten most of what used to be hand-built. The early self-driving systems had been modular, a chain of separate components each tuned by engineers, one to find the lane lines, one to detect the cars, one to plan the path, each handing its output to the next. Version by version those hand-tuned stages gave way to a single network trained from end to end, fed the raw camera images at one side and producing the steering and the pedal at the other, having grown for itself the intermediate understanding the old modules used to be programmed to provide. The hand-written rules retreated to the edges, and the grown network moved to the center of the car.

A case is born from a moment of trouble. Most often it is a disengagement, the instant a human driver loses faith and grabs the wheel or touches the brake, and the system notes the takeover, marks the seconds of footage around it, and files the moment as an example of a situation it handled poorly enough that a person felt the need to step in. Other cases come from the system flagging its own uncertainty, the merge it completed with low confidence, the object it could not classify cleanly. Each one is a few seconds of the world seen through eight cameras, a small parcel of exactly the hard situation the next network most needs to study. The fleet, with nobody driving it for that purpose, is forever sorting the world into the parts the machine already handles and the parts it does not, and shipping the second kind home.

The reason the loop needed a fleet of that size was the shape of the problem. The ordinary cases of driving, the empty road and the clear intersection and the gentle highway curve, a network learns in its first weeks from its first thousand drivers. What remains after that is a long tail of the rare, the situations any single car meets once in fifty thousand miles. A mattress fallen off a truck into lane two. A traffic officer waving cars through a dead signal against the light. A flooded dip in the road at night with the lane lines under water. Each of these is individually rare, and there are millions of distinct kinds of them, and the only way to gather enough examples of enough of them to grow a network that handles them is to be driving tens of millions of miles a day. A small fleet never meets the tail. It drives its thousand careful miles and sees the same easy situations over and over, and the rare events that decide whether autonomy can be trusted never arrive in its data at all.

Some of the tail could be met without waiting for it. Once the fleet had captured a rare event a single time, the engineers could rebuild that moment in simulation and run the car through it again and again, varying the speed of the fallen mattress, the angle of the low sun, the moment the child darts out, generating thousands of versions of a situation the real world had offered only once. The network could then be grown against the whole family of a hazard, every variant of it, where before it had only the single instance the road happened to provide. The real fleet found the rare thing. The rebuilt world multiplied it into enough examples to learn from. Between them they reached into the long tail in a way that driving alone, however many miles, would have taken decades to manage.

Some of the most useful data came from perfectly uneventful driving. A Tesla running in what the engineers call shadow mode keeps the network thinking in the background while a human drives, computing what it would have done at every moment and comparing that against what the human actually did. Where the two agree, there is nothing to learn. Where they diverge, where the network would have braked and the human did not, or the network would have held and the human went, the disagreement is itself a flag worth uploading. The fleet could try a new behavior across millions of cars in silence, watching how often it would have matched a careful human, before that behavior was ever allowed to touch the controls. The cars were running the experiment while their owners ran their errands.

Miles per intervention, the average distance the car drove before a human felt the need to take over, climbed release after release. Each new version, grown against the hardest cases the last version had collected, went further between takeovers than the one before it, and the curve of that number over the years was the curve of the flywheel turning. When the figure stalled, it told them the data they were feeding the network had stopped containing the situations it still got wrong, and the work became finding those situations in the fleet. When it climbed, it told them the loop had found something new to learn from and learned it.

Tesla's engineers can turn those flagged moments into the next network's training and tests. Fleet size gives the team a chance to see a strange crossing or an uncertain cyclist more than once, on roads the engineers did not choose. A candidate release can be run through the Mission turn alongside thousands of older cases; the team can see whether a better gap judgment came at the cost of a worse response somewhere else. The fleet brings in the hard moments. The engineers decide what the next network has actually learned.

Every hard case in the library had begun as a small human loss of faith, a driver who flinched, who grabbed the wheel, who decided in some ordinary second that the machine was about to get it wrong. The system grew most from exactly the moments people had refused to let it continue. Each of those refusals was filed, studied, and turned into training, so that the next version would handle the thing the last one had fumbled, and the human in that seat would have one less reason to reach for the wheel. The machine was being taught, case by case, by the very mistrust it was working to retire.

The loop kept turning. Drivers found the edge of the current network, engineers made those moments into tests, and a revised network went back out to meet traffic again. The difficult part of the loop was the ordinary street that would not repeat itself on command. A correction found in one car could reach many others before their drivers faced the same unusual crossing. The fleet could be a way of sharing a learned piece of road judgment across places that had never shared a road.

Taking the hand away from the wheel means teaching the system to meet rare situations on real streets, then testing each release against them. In San Francisco, the safety driver still sits within reach of the controls. If the system reaches that point, a person whose eyesight no longer lets her drive could make an ordinary cross-town trip on her own schedule. The work needed for that trip is present in the supervised turn, still being measured.

THE TURN

Back at the Mission intersection the supervised Robotaxi eases its nose toward the centerline. The safety driver watches the next gap open, ready to intervene while the system reads the pedestrian, the cyclist, the clumps of cars and the double-parked truck ahead. The car takes the turn cleanly. Millions of filed corrections sit inside that motion, and the car settles into the cross street under the driver's eye.

9

Two Oceans of Data

THE DEER AT THE VERGE

ON A COUNTY ROAD in eastern Tennessee with nothing else on it, a Tesla running on Autopilot registered a shape at the right edge of the road and began, very slightly, to slow. It was fourteen minutes past two in the morning. The shape was a white-tailed deer, standing at the verge with its head turned toward the oncoming headlights, and the car's perception network had picked it up and marked the right lane as partly compromised before the driver, who was tired and only half watching the road, had seen anything. The car eased off. The deer made up its mind and crossed, unhurried, and was gone into the dark on the far side. The car came back up to speed. The whole event lasted perhaps four seconds, and the driver, if asked the next morning, would not have remembered it, because nothing had happened.

The encounter could become a training case only after one decision onboard: which few seconds were worth sending home.

What went back over the network was tiny. A modern car generates far more data in an hour of driving than any network could afford to ship home, terabytes of raw video from eight cameras, and almost all of it is the ordinary world rolling past, empty lane and clear sky and the backs of other cars, worth nothing to a system that already knows how to drive a clear road. Sending all of it home was never possible and was never the point. The car kept its footage locally, ran the flagging logic onboard, and uploaded only the handful of seconds around the events its own software judged interesting, the near-miss, the strange object, the moment its confidence dipped, the disagreement between what it did and what the driver did. The deer at the verge was a few hundred kilobytes wrapped in a parcel of metadata, the speed and the location and the time of night and the version of the software that had been running. Out of an hour of driving, a few seconds were worth keeping, and the car's first job, before it was a vehicle at all, was to know the difference.

The moments that improve them are the moments nobody notices, the near-misses that resolve into nothing, the small hesitations a passenger would never feel and a driver would never recall. A human driver gets better through the events that frighten him, the close calls he remembers for years afterward. The fleet got better through the close calls that no one remembered at all, gathered by the million, each one too small to register on anyone in the car, each one a tiny correction sent home. The learning happened in the gaps of human attention.

THE FLEET AS INSTRUMENT

To the people who run the pipeline that the Tennessee event flowed into, the remarkable thing is the volume. Ashok Elluswamy runs Autopilot and full self-driving AI at Tesla. His team helps turn encounters like the deer into training examples drawn from one of the largest deployed camera fleets in the world. By 2026 the fleet numbered several million cars. Each one carries eight cameras looking out in every direction, motion sensors, satellite positioning, and enough onboard computing to run the perception network in real time on all of it as the car drives. Each car, every day it is driven, is taking the measure of the world.

The fleet had become a new scientific instrument. The great instruments of the past had each been built to measure one thing very well. A telescope gathers light from the sky. A particle collider watches matter break apart. Each is enormous, fabulously expensive, aimed at a single question, and run by a small priesthood of specialists. The fleet measured nearly everything that happens on a road, everywhere roads exist, all the time, and several million customers had paid for it themselves, each buying a car for reasons of his own. No government had funded it. No laboratory had planned it. It had assembled itself, one purchase at a time, into the most comprehensive ongoing measurement of human movement ever made, and the company that received its readings had only to build the machinery to make sense of them.

From where Elluswamy's teams sit, the Tennessee deer barely registers as an individual thing at all. It is one row among millions arriving that night, a single entry in a flood of flagged events that streams in from every time zone at once, so that the queue is never empty and the sun is always rising on some part of the fleet. The work is to build the machinery that can take in millions of these a day, sort the useful from the merely odd, and turn the useful into the next version of the network without a human ever having to look at most of them. A person sitting and watching the raw feed scroll past would learn nothing. The volume is far past what any person can read. The craft is entirely in the sorting.

The fleet drives in the snow in Norway and the monsoon in Mumbai and the stop-and-go of São Paulo and the empty county roads of Tennessee at two in the morning, in fog and glare and construction and the particular chaos of a school pickup line, on six continents, in conditions no engineering team could stage on purpose if it tried for a hundred years. What flows back from all of that, filtered down to the moments the network found hard, is a stream of training material whose scale has no comparison anywhere else in the industry.

Raw events are worth little until they are labeled. A clip of a deer at the verge teaches a network nothing until the deer is marked as a deer, its position fixed in three dimensions, its motion traced frame by frame, the road edge and the lane lines and the other objects in the scene all named and located, so the network can be told, in effect, this is what was there and this is what you should have understood. For years that labeling was done by hand, by rooms full of people drawing boxes around objects in still images, and at the scale of a global fleet hand-labeling everything was hopeless from the start. The pipeline's answer was to automate the labeling. A very large network, too slow and too expensive to ever run inside a moving car, is run later in the data center against the uploaded clips, and it labels them automatically, marking the objects and tracing their motion far more thoroughly than the car had time to in the moment. The people no longer draw the boxes. They audit the machine's boxes, spot-checking its work, correcting it where it errs, feeding the corrections back so the automatic labeler grows more accurate too. Labeling had itself become a learned task that improved the more of the world it was shown.

Choosing what to keep was harder than labeling a clip. A fleet of several million cars could flag far more events than any system could ever train on, and most flagged events, once examined, turned out to be redundant, the thousandth example of a merge the network already handled in its sleep. Keeping all of them would have buried the rare and valuable cases under mountains of the common and useless. So the pipeline learned to prize scarcity. It hunted for the events that were unlike the events it already held, the situation seen only a handful of times across the entire fleet, and it weighted the training toward those. The most valuable data in the building was the data there was least of. Curating the stream, deciding what earned a place in the library and what could be thrown away, was the part of the work no machine could yet be trusted to do alone, and it was where the engineers earned their keep.

In the early years people decided what the network needed and went looking for it. Later the network began to say so itself. A model that had been trained could be run against fresh footage and asked where it was unsure, where its predictions wavered, where two plausible readings of a scene sat too close together for comfort. Those moments of doubt were exactly the footage worth keeping, because they marked the edges of what the system had already learned. So the fleet could be sent looking for them. A network uncertain about cyclists at dusk could ask, across millions of cars, for more cyclists at dusk, and the cars would return them over the following days. The data engine had become a thing that knew its own gaps and reached out to fill them. The model had begun to nominate its own next lessons, and the people who had once chosen them now mostly checked the choices. It was a strange inversion, a student writing the next page of its own syllabus, and it made each turn of the training cycle sharper than the last.

Turning the curated library into a better car was its own long discipline. A new version of the network was trained against the whole accumulated library, the millions of kept events, and then it had to be proven safe before a single line of it reached a public road. The danger was always that fixing one thing broke another, that a release taught to handle the deer at the verge would somehow drive worse at a crowded four-way stop it had managed fine the week before. So every candidate version was run back against an enormous bank of held-aside situations, scored on how it would have behaved in each, and compared release against release across tens of thousands of scenarios at once before anyone judged it ready. Only after it cleared that gauntlet did it go out over the air to the fleet, downloaded overnight while the cars charged, so that millions of vehicles woke a little more capable than they had gone to sleep. The cycle from a flagged event to a fix in every car took on the order of weeks, and it never stopped turning.

All of this had to live somewhere. The clips and their labels and the trained networks and the held-aside test banks added up to a quantity of data that strained the ordinary meaning of the word *large*, measured in the millions of gigabytes and climbing every day the fleet drove. It sat in data centers full of spinning disks and flash memory, cooled and powered and backed up, a physical library the size of a warehouse district holding a compressed record of how the world moves. Storing it was a serious cost in its own right, and part of the curation discipline was knowing what could be discarded to keep the bill from growing without bound. It took buildings and electricity and machines to hold, and the holding was a price the company paid every single day, against the bet that what it held was worth more than the cost of keeping it.

There was a question of whose data this was, and it sat uneasily under the whole arrangement. The cars belonged to their owners, the trips were the owners' trips, the streets were public, and the clips that some cars shared were recorded by cameras pointed at a world full of people who had never agreed to be measured. Tesla's privacy notice described the vehicle data it collected; sharing camera clips for fleet learning required a driver's consent in the car's settings. The legal terms described that exchange, but did not settle the larger question. The deeper question, of what it means for a single company to hold visual samples of roads across the world, gathered by machines their owners had paid for and drove for their own reasons, remained open in 2026. The data kept arriving while the question stayed open.

THE X FIREHOSE

There was a second ocean, and it was made of language. In 2022 Musk had bought Twitter, the global town square that became X, and for a long time the public argument about that purchase was about speech and moderation and money. Underneath that argument was a quieter fact about what X actually is, which is one of the largest continuously generated streams of human text and reaction in existence, hundreds of millions of people writing to each other in real time about everything that happens as it happens. The artificial intelligence company Musk had founded had taken in X itself the year before, and in February of 2026 that combined entity was folded into SpaceX, bringing X and the model-building team into the company that made rockets and Starlink satellites. Tesla remained separate, with its road fleet and driving models.

A platform like X receives language at the moment people make it. Someone reports a bridge closure in a phrase a local driver would use; others reply with a detour, a correction, or a photograph. The first post may help answer a question about the road today if it can be checked and retrieved. Later, selected examples can teach a model how a place is named and how an answer is corrected. A social stream can supply both the changing facts people ask about and examples of the language in which they ask.

By the time X entered SpaceX, the open supply of human-written text had become a resource the labs could no longer treat as inexhaustible. The frontier labs had spent the better part of a decade training on the public internet, on the books and the articles and the forum posts and the code that people had put into the world, and the same words could not serve forever as the only source of new learning. Fresh text, written by real people about things happening now, was valuable for that reason. A platform where hundreds of millions of people wrote to each other in real time, reacting to the news of the hour in dozens of languages, was a renewing spring of exactly that. The posts were one part of it. The reactions were another, the replies and the reposts and the small signals of approval and disgust that recorded how the world had received each thing that was said. X produced on the order of three hundred and fifty million posts a day, most of them fresh samples of human language and human judgment, generated for free by people doing what they would have done anyway.

Only part of the firehose was worth drinking. A great deal of what flows across a social platform is noise, spam and slogans and bots talking to bots, the same empty phrases repeated a million times, and a model trained naively on the raw stream would learn the internet's worst habits along with its knowledge. The value lived in the filtering, in pulling the genuine human writing and the real human reactions out of the torrent and leaving the rest behind. There was also a live edge to the platform that static, scraped data could never have. When Grok answered a question in front of the platform's users, those users replied to it, argued with it, corrected it, praised it, and their responses were themselves a signal about how good the answer had been, available to the people training the next version. The model met its public in the open, and the meeting was recorded.

The timing matters. Training changes the weights of a later version of a model, after examples have been chosen and tested. A live query can be answered sooner through retrieval, provided the report is found, dated and checked against other sources. During a storm, a person asking which road remains open needs the present tense, not a remembered map from last year's training. The platform can bring that present tense within reach, while the filtering determines whether the answer deserves to be trusted. It is the difference between knowing and having known.

The road fleet and the language platform gathered different kinds of evidence. A driving clip shows the movement a car had to judge; a post and its replies show how people describe a new event and contest an answer. Tesla held the road footage and SpaceX held the X stream. Imagine a later car that can explain in speech why it stopped for a deer it saw before its passenger did. That service would need the visual judgment to act safely and the language skill to answer the passenger's question accurately. The work of combining them lies between the two libraries.

The deer clip can be used again. It can test this year's network, help train the next one, and be compared with a new set of night-road examples without being consumed. Its worth still depends on accurate labels and on roads, vehicles and rules that may change. The language stream is more perishable; a report about today's closure may be useful tonight and historical context tomorrow. These libraries grow valuable through keeping, selecting and renewing, not simply through accumulating everything that passes through them.

SYNTHETIC, SIMULATED, SAMPLED

Real data, even at the scale of a global car fleet and a global text platform, ran out. There are behaviors a model needs to learn that the world simply does not supply enough clean examples of, rare events, dangerous situations no one would stage, the long tail of cases that occur once in ten million miles and that a safety-critical system has to handle the first time it ever sees them. To cover that ground, model builders manufactured data, and Tesla used simulation in its driving work. Other models generated training examples. Physics engines simulated the dangerous situations that could not be allowed to happen for real. Procedural systems rendered millions of variations of a scene, the same intersection under every light and weather and traffic condition that a single real intersection would take a decade to actually experience.

Manufacturing data carried its own danger, and the field had a name for it. Model collapse. A system trained too heavily on the output of other systems can drift away from the real world it is meant to describe, learning the habits and the blind spots of its teachers and amplifying them with each generation, the way a photograph of a photograph loses something each time it is copied. The synthetic data had to be anchored, checked against the real, kept honest by road cases for a driving model and human writing for a language model. The world still kept the models from disappearing into their own reflection. This was the reason no lab could simply dispense with real data and generate everything it needed. The captured and the manufactured had to be held in balance, and getting that balance wrong in either direction cost a lab its progress.

The proportion of manufactured data to captured data changed from one training run to the next, and somewhere in the middle of all this the people building the models noticed something about where their progress was coming from. The frontier of what a model could do was moving with the quality of the synthetic data more than with the sheer volume of the real data. The world was the seed. The simulator was becoming the field where the crop was actually grown.

Manufactured data had one quiet advantage the real world could never offer. When a scene is generated, its internal facts are known exactly, because the system that made it knows where it placed every object, every speed, every distance, down to the last detail. The generator writes an answer key as it builds; engineers still have to audit what that key is worth. Real data has to be interpreted, and the interpreting is slow and costly and never perfectly right. Synthetic data skips the first pass of annotation. As the tools for generating it improved, the cost of producing a million labeled examples of a hazard fell toward the cost of the electricity needed to compute them, and a lab that had once been limited by how much of the world it could afford to capture and label found its new limit in how much computation it could bring to bear on manufacturing worlds of its own.

The manufactured world carried its own risk. A simulated world is only as truthful as the physics and the rendering that built it, and a model can learn to handle a hazard beautifully in simulation and still be caught out by the real version, which is always stranger and grainier and less obliging than anything a renderer produces. The gap between the simulated and the real was a thing the engineers spent enormous effort measuring and closing, because a lesson learned in a world that is slightly wrong is a lesson that may have to be unlearned the first time the real road declines to behave like the model.

By 2026 the work of building machine intelligence was fed by three rivers. Tesla's fleet poured in the physical world, a daily measure of roads in many countries. X poured in the human one, the language and judgment of hundreds of millions across the connected world. Simulators poured in worlds of their own making, the rare and the dangerous and the merely useful, manufactured to order. All three fed libraries of examples built for different models, and for a few years the size and richness of those libraries had decided who led and who trailed. That era was ending. The libraries were full and still filling. What now decided the pace of progress was the computing engine that could draw them down and turn them into trained intelligence, and that engine ran on something their builders could not gather or generate or buy in the quantity they needed.

THE CONSTRAINT MOVES TO THE ROCK

A decade of building frontier AI had taught the field one durable lesson about its own progress, which is that the thing in shortest supply never stays the same thing for long. Early on, when training data was scarce, the models were kept small because there was nothing to feed a larger one. When techniques arrived for building much larger models, the hunger of those models made data the scarce thing. When the fleet and the firehose and the simulators eased the data problem, the scarce thing became the computation, the sheer quantity of computing needed to train a frontier system. And by 2026, with the labs awash in data and pouring money into computation as fast as it could be installed, the scarcity had moved one layer deeper and turned physical. There were not enough of the right chips, and beneath that shortage lay a deeper one, the small number of places on Earth able to manufacture such chips at all. The constraint now lived in the silicon, in physical wafers etched in physical buildings, and there were only so many of those.

The fleet could find another hard case that night. Expanding the computing that trained on those cases depended on wafers made in another company's foundry, on a slower schedule than the cars collected footage.

10

Chips as Choke Point

THE ISLAND

ROUGHLY A HUNDRED and fifty kilometers off the southeastern coast of mainland China there is an island a little larger than Belgium, mountainous down its spine and crowded along a narrow western plain, with some twenty-three million people living on it. On that plain, in an industrial park outside the city of Hsinchu and at a handful of sister sites scattered down the island, stand what are arguably the most consequential factories on Earth. They belong to the Taiwan Semiconductor Manufacturing Company, and what comes out of them is the patterned silicon on which nearly everything else now runs.

The concentration has a number attached to it. On the order of ninety percent of the world's most advanced logic chips, the leading-edge processors that sit at the front of the line, are fabricated on that one island, the large majority of them by that one company. Apple's phones, Nvidia's accelerators, AMD's processors, the radios from Qualcomm and MediaTek, and the entire wave of artificial-intelligence silicon that came to define the decade from 2024 onward are designed in California and Texas and Seoul and turned into physical chips there. The firms that design the chips do not, for the most part, build them. They send the designs to the island.

The arrangement was an invention, and a recent one. For the first decades of the industry a chip company built what it designed, in its own factories, the two halves of the business inseparable, because designing chips and making them were thought to be a single skill. What pulled them apart was the cost of the factory. As each new generation of fab grew more expensive than the last, climbing from hundreds of millions of dollars toward tens of billions, fewer and fewer design firms could afford to own one, and a company on the island made a wager that it could live by building chips for everyone and designing none of its own, taking the orders the design houses could no longer afford to fill themselves. The wager was TSMC. It let a designer in California exist on the strength of its ideas alone, with no factory at all, sending its blueprints to a fabricator that would never turn around and compete with it for customers. The arrangement freed a generation of chip designers from the crushing cost of a fab, and by the very same motion it gathered the making of the world's chips onto the few foundries with the nerve and the capital to keep building them.

In 2014 IBM agreed to pay GlobalFoundries roughly a billion and a half dollars to take its chip-manufacturing business away. The transfer closed in 2015. IBM had paid to be rid of making the very chips it still designed. Apple has never owned a fab at all, and only a few firms still keep both the design and the fabrication under one roof. The manufacturing half of the business had grown heavy enough that owning the factory, at the leading edge, became a liability a company would pay to escape.

To call a chip advanced is to make a claim about size. The features etched into the silicon, the individual transistors and the wires that join them, had been shrinking for sixty years, each generation packing more of them into the same area and switching them faster and at lower power, and by the middle of the 2020s the smallest of those features had fallen to a few billionths of a meter, narrower than many viruses, approaching the width of a handful of atoms set side by side. Drawing patterns that small onto silicon needs light of an extraordinarily short wavelength, and bending that light to the task needs a machine whose complexity has few equals in any industry. There is exactly one company on Earth that makes the machines that print the most advanced chips, a Dutch firm called ASML, and a single one of its newest systems costs on the order of a few hundred million dollars, ships in hundreds of crates, and takes months to install and tune. The island's dominance sits on top of that machine. TSMC buys the Dutch machines and runs them at higher volume and higher yield than anyone else alive, which means the choke point everyone worried about had a second choke point standing above it, one company making the tools and one company, on one island, using them best.

C.C. Wei, who runs TSMC, presides over a manufacturing operation with no real second anywhere in the world. A competitor that wanted to take even a slice of that business could not do it by deciding to. The plants represent decades of process knowledge, tens of billions of dollars of equipment, and a workforce trained to tolerances measured in the spacing of individual atoms, and there are only a few places on the planet that hold any part of that capability at the leading edge, with the island holding most of it. The most important industrial concentration in the modern economy sits on a single contested island, within artillery range of a far larger state that claims it as its own. Everyone who designs a chip knows this. For most of the 2020s the knowledge sat in the background of every roadmap like a fault line under a city, mapped, understood, and built on anyway, because there was nowhere else to build.

This dependence was thought to give Taiwan a silicon shield. The reasoning held that no great power would risk destroying or seizing the one place that made the chips the world economy now ran on, because the harm would rebound on the aggressor and on everyone else within months, as the supply of advanced logic simply stopped. Whether a shield made of economic consequence actually deters anyone was a question no one wanted to test, and the planners who thought about it for a living were not reassured. A blockade well short of invasion, a quarantine that merely interrupted the flow of chips off the island for a season, would be enough to idle assembly lines for cars and phones and data-center servers across three continents, and there was no stockpile deep enough and no second source large enough to cover the gap. The world had arranged its most important industry as a single point of failure and then talked itself into treating the fragility as safety.

C.C. Wei, an engineer who had risen through TSMC's process side, led the company through this expansion. Arizona, Japan, and Germany answered customers and governments that wanted capacity in more than one place. Yet the newest processes still had to be developed, tuned and made reliable at scale, and Taiwan held the thickest concentration of people, tools and suppliers who knew how. An overseas fab could be built faster than that accumulated learning system could be moved. The geography of the frontier changed more slowly than the map of its factories.

THE ARIZONA TEST

The obvious answer to a concentration that dangerous is to spread it out, and the obvious place to spread it was the United States. The story of the attempt is by now familiar. In 2020 TSMC announced that it would build advanced fabrication plants in the desert north of Phoenix, Arizona, the first serious move of leading-edge capacity off the island in a generation. The plan ran on a timeline of about four years to first production. The first Arizona plant began high-volume production of four-nanometre chips in the fourth quarter of 2024, roughly on that original four-year horizon. Getting there had still meant construction and equipment delays and a public labor dispute when the company found that the technicians it needed, people trained to the island's exacting standards, were scarce on the ground in Arizona and had in some cases to be flown in from Taiwan to train the local crews. By the time the first Arizona fab was turning out chips, the node was advanced, even as Taiwan remained the site of the company's newest process.

A generation behind, in this business, is a different business. The most advanced node commands the premium, the high prices the newest phones and the hottest accelerators will pay to be first and fastest, and that premium is where a fab earns back the tens of billions it cost to build. A node a generation old has already slid past the frontier's premium, doing honest work at thinner margins for buyers who have no need of the bleeding edge. So an Arizona plant that opened a generation behind the island opened, in a sense, into the wrong market, making yesterday's chips at today's costs while the leading edge, and the money that chases it, stayed home on the island. The plant proved the United States could fabricate advanced chips on its own soil. It could not yet prove the country could do so at the very front of the line, and the front of the line was the only place the strategic worry had ever really lived.

The friction went deeper than the schedule. The Taiwanese engineers TSMC flew to Arizona to stand up the plant arrived with a working culture built around long hours, exacting discipline, and an assumption that the fab came before nearly everything else in a person's life, and they met an American workforce that did not see the job that way and had no intention of starting to. There were disputes over the hours, over the chain of command, over the very definition of what counted as showing up. Experienced American technicians left. Some of the Taiwanese staff, posted far from home to a desert they had not chosen, wanted to go back. Money could not compress a problem made of habits and expectations, and habits take as long to rebuild as they took to form. The plant did run. Its first high-volume production reached the initial 2024 horizon, but the route there showed how much of the island's working culture had to be learned alongside the machines.

Underneath those frictions was a single fact about what a fab actually is. The thing that could not be shipped to Arizona was not the building or the lithography machines, which money can buy and a heavy crane can set down on any poured slab, but the density that had grown up around the island sites over forty years, the hundreds of specialized suppliers within an hour's drive, the technicians who had spent entire careers inside one process, the accumulated and largely unwritten knowledge of how to coax a high yield out of a temperamental line. That density cannot be ordered and installed. Growing it costs the one input that no quantity of capital compresses, which is time. What the Arizona project demonstrated, in the most expensive way available, is that a local manufacturing ecosystem has to be cultivated beyond the first high-volume wafers, with suppliers and experienced teams growing around the line over years.

The word the industry uses for what that density buys is yield. A fab does not turn every wafer it starts into working chips. Some share of the chips on every silicon disc come out dead, killed by a speck of dust or a fractional misalignment or a thousand subtler faults, and the fraction that survive is the yield, and the yield is the difference between a plant that prints money and one that bleeds it. A new line might begin its life yielding well under half its chips and climb, over months and years of relentless adjustment, toward the high numbers that make the economics work at all. Every point of that climb is paid for in the experience of people who have watched the line fail in every way it can fail and learned to keep it from failing the same way twice. That learning is the density. It lives in the hands and the habits of the workforce and in the nearness of the suppliers who can fix a misbehaving tool the same afternoon it breaks, and it is exactly what a new plant in a new country does not have, cannot buy, and has to earn the slow way, one ruined wafer at a time.

THE FLOOR UNDER THE FLOOR

Every layer of a modern industrial system comes back, in the end, to silicon. The energy layer runs on power electronics switched by chips. Transport runs on the inference processors that read the road. The factory runs on the controllers that drive the machines down the line. The robot runs on the chip that turns its model into motion. And intelligence itself, the models that had begun to sit underneath all the rest, is computation performed at enormous scale on silicon built for the purpose, and nothing else. The bottleneck that had surfaced at every level of the industrial stack so far, where one input falls short and sets the pace for everything stacked above it, reaches the chip and keeps going. It runs down into the question of who can fabricate the chip, and where, and on whose schedule. A company can own its mines and its refineries and its assembly lines, can design its own chips down to the last transistor, and still find that the physical making of those chips rests with someone else, on ground it does not control and on a schedule it does not set. A firm that has done everything else for itself and still rents the etching of its silicon keeps one part of its loop in another company's hands, the most expensive and least movable part there is. The company that controls its own silicon sets its own pace, its own roadmap, and its own ceiling. The company that buys its silicon takes the pace its supplier is willing to give it, and discovers the real terms of that arrangement only in the year when the supplier has more buyers than wafers.

In 2020 and 2021 the whole argument got tested in the open. Pandemic buying collided with chip capacity that sat in a few fabs, and Ford and General Motors and Toyota and most of the rest idled their assembly lines for want of ordinary microcontrollers, the cheap commodity chips scattered by the hundred through a modern car. Something like eleven million vehicles went unbuilt and the industry lost well over two hundred billion dollars, and the part that stopped the line was the cheapest silicon in the vehicle, not the exotic processor. Tesla was the conspicuous exception, because it wrote its own vehicle firmware. Controlling that software boundary, it could rewrite the code around whatever substitute chips it could actually buy, qualifying on the order of nineteen new controller variants while the crunch ran, and it kept building cars while the companies that had outsourced their firmware waited on their suppliers. Owning the software layer turned a supply shock into a schedule the company still set for itself.

THE COMPUTE TESLA ACTUALLY BURNS

While the public argument about Tesla and chips was still about whether the company could ever design its own, Tesla had assembled one of the largest computing clusters in private hands and pointed all of it at the driving problem. The cluster has a name, Cortex, and it sits inside the company's plant outside Austin. By early 2025 it held on the order of fifty thousand of Nvidia's H100 graphics processors in a single facility, wired together to train the autonomy models and, alongside them, the models that would run the humanoid robot, and it was scaled further through 2026 as those two training loads grew. Tesla had built it out of chips bought from Nvidia, the same chips every other lab on Earth was bidding for, because the chips it would rather have used did not yet exist.

A cluster like Cortex is easier to name than to picture. Fifty thousand high-end processors do not sit on a desk. They fill a purpose-built hall with row upon row of racks, each rack a wall of computers, all of it drawing power on the scale of a small town and turning very nearly all of that power into heat that has to be carried off as fast as it is made, by chilled water and moving air, around the clock, or the machines cook themselves in minutes. At full load, the electricity alone could run into the tens of millions of dollars a year, and the building draws on the grid like a steel mill. This is the part of artificial intelligence that never shows up in the demonstrations, the physical plant of it, the substations and the cooling towers and the generators waiting for the night the grid falters. The intelligence that felt weightless to the person typing at it was, at the other end of the wire, one of the most power-hungry machines ever assembled, and every further increment of it had to be paid for in megawatts.

The H100 was a single graphics processor made by Nvidia, about the size of a hardcover book, and through the middle of the decade it had become the most fought-over manufactured object in the world. It was the standard unit of artificial-intelligence computing, the thing you counted to say how much intelligence you could afford to train, and every serious laboratory wanted as many as it could get. The supply was fixed by how many Nvidia could have fabricated on the island, and in 2024 and 2025 delivery slots were scarce and fiercely contested. A buyer with deep pockets and a year's patience could assemble a cluster. A buyer the size of Tesla, wanting tens of thousands of them and wanting them at once, was bidding against every other giant for a supply that could not grow as fast as the hunger for it. A single chip cost into the tens of thousands of dollars, the largest buyers were spending billions a year, and all of that money ran to one designer in California and one fabricator on one island. To own the chip you trained on, if you could manage it, was to step out of that line.

Building your own training chip is one of the hardest things a company can attempt, and the reasons to try anyway come down to cost. A chip designed for one company's exact workload, stripped of everything that workload never uses, can in principle do more computing per watt of electricity and per dollar of capital than a general part built to satisfy everyone at once. At the scale of a company training autonomy models and robot models around the clock, a chip even somewhat more efficient than the one it was renting from the market would pay back its enormous development cost and keep paying, year after year, in lower power bills and faster training and freedom from the allocation line. That was the prize. The catch was that reaching it meant matching, from a standing start, a competitor that had spent decades and tens of billions of dollars learning to do the one thing, and matching it while it kept moving. Most companies that tried this failed. The few that succeeded, all of them among the largest in the world, found that the chip they ended up with was worth more to them than almost anything else they owned.

Tesla reached that conclusion as part of a larger turn. By the middle of the decade every one of the handful of companies large enough to afford the attempt was designing silicon of its own. The big search company had been running its models on custom chips of its own design for years. The largest cloud providers had each built their own, one kind for training the models and another for the everyday work of serving them to customers, expressly to stop handing so much of their revenue to a single supplier. The pattern was the same in every case. A company reached a scale at which the chips it rented had become its single largest cost and its sharpest constraint, and at that scale the brutal economics of designing its own started, for the first time, to add up. What set Tesla's version apart was the second half of the loop, the fleet out on the road, which meant its chip had to be cheap and rugged enough to ride in a car and strong enough to help train its own successor, the same part working both ends of the same problem.

The attempt to make those chips had a name too, and a history worth walking. Dojo was the in-house silicon program Tesla announced in 2021, built around a custom training chip the company called the D1, and its entire purpose was to reduce the company's dependence on Nvidia by training its networks on hardware of its own design. In August 2025 Tesla disbanded much of the original Dojo team and shifted its immediate focus to AI6. By 2026 Tesla was developing Dojo 3 as another generation of custom training silicon. Ganesh Venkataramanan, the engineer who had built the first Dojo chip before leaving Tesla in 2023, was already gone, and at the shutdown a group of the remaining Dojo engineers left to pursue the architecture in a company of their own, with Venkataramanan at its head. The chip problem did not disappear with the original team. AI6 and the later Dojo 3 effort pursued it through a changed architecture.

Musk's framing at the time was that every path the company had explored had collapsed onto a single chip, AI6, and that Dojo's second-generation design had become an evolutionary dead end not worth feeding while AI6 existed. The way a decision like that got reached inside the company had a name on the factory floor. They called it walking to the red. In a conference room at Fremont named Jupiter, monitors showed the status of the line floor by floor and station by station, green where a station was keeping pace and red where it was not, and Musk would walk the floor toward whatever was red, with a heavy bear-like gait, making something like a hundred command decisions a day about what to change, what to cut, and what to keep. He knew some of those decisions would be wrong, perhaps one in five. "But if I don't make decisions, we die," he said. The decision to end Dojo 2 belonged to that same willingness to discard an architecture after years of work. Its eventual cost or wisdom could only be judged later.

AI6 was the chip Tesla actually wanted, and the company placed it at an existing foundry without waiting to build one of its own. The contract went to Samsung, for fabrication at Samsung's plant in Taylor, Texas, a deal worth sixteen and a half billion dollars signed in July of 2025. Tesla would design the chip and Samsung would build it, and the chip was designed to run in both halves of the same loop, inside the cars where it read the road in real time and inside the data centers where it trained the next version of the network that read the road. The same chip in both halves of the loop. Musk announced the arrangement himself, with the specifications laid out on an earnings call, because the chip had become central enough to the company's story that the founder carried the news. The organization that has to turn the AI6 design into something Samsung can actually etch is run by Anant Nivarti, who leads Tesla's silicon engineering across the AI6 work and the training hardware behind it.

The detail worth pausing on is the single architecture running in both places. For most of the industry the chip that trains a model in a data center and the chip that runs the finished model out in the world are separate designs from separate companies, built to opposite priorities, the training part huge and power-hungry and the deployed part small and frugal. Tesla's wager was to make one design serve both, the same silicon in the car reading the road and in the data center learning from everything the cars had seen. A chip that does both can be manufactured in far greater volume than either job alone would justify, which drives its unit cost down, and it lets the thing that learns and the thing that acts speak one native language, so that an improvement trained on the cluster drops onto the car without translation. Whether a single design could be genuinely excellent at two jobs that usually pull in opposite directions was the open question the whole gamble turned on. If it worked, Tesla would be building, in volume, the one chip that closed its own loop. If it failed, the company would have spent years and a fortune learning what the rest of the industry already believed, that the two jobs want two chips.

AI6 offered Tesla a route out of Nvidia's allocation line, but not yet a chip it could train on. Samsung would have to build it in Texas, on Samsung's equipment and schedule and at the yield Samsung could reach. The design was Tesla's own and the volume committed years ahead; those commitments could give it more control than buying finished chips on the open market. The dependency would then shift one box to the left, from chip designer to fabricator.

TERAFAB

The reflex that would eventually produce Terafab had been at work inside Tesla for years. As Tesla planned a Nevada battery factory, Panasonic hesitated to commit. Musk and JB Straubel had work lights and bulldozers sent to the proposed site near Reno, then invited a Panasonic representative to see the ground being prepared. Later, over dinner in Japan, Musk made the case to Panasonic president Kazuhiro Tsuga for building the factory together. Tsuga agreed that Panasonic would finance roughly two billion dollars of the projected five-billion-dollar factory. Panasonic would make the cells; Tesla would build them into packs. The partnership secured the supply Tesla could not yet make alone, and taught the company how much of its pace still lived in another firm's hands. When a supplier sets your pace, you eventually build the layer yourself.

The Terafab idea reached below chip design and into the line that would make it. SpaceX and Tesla wanted computing hardware at a scale their existing suppliers could not deliver. The first step would be a design a foundry could print, a process that could repeat it, and enough good dies on each wafer to make the line worth running.

Tesla had experience designing chips for a car's immediate decisions and for the larger machines that train them. Intel held the process knowledge and packaging equipment needed to turn designs into working silicon in volume. A wafer would pass through many patterned steps before testing found the dies that actually worked; packaging would then have to connect those dies to the memory, power and cooling the finished machine needed. The suppliers of chemicals, tools and parts would have to reach the line when a step called for them. The first useful measure of the project would be a repeatable working part, not the size of the drawing that preceded it.

There was a reason Intel was available to be asked. For most of the history of the industry Intel had been its center of gravity, the company whose name sat inside nearly every personal computer and whose factories set the pace the rest of the world measured itself against. Through the late 2010s and into the 2020s it stumbled, missed a sequence of manufacturing transitions, watched the island pull ahead, and arrived in the middle of the decade holding extraordinary process knowledge and not enough customers to fill the fabs that knowledge had built. A government alarmed by the concentration on the island had begun pouring money into rebuilding American fabrication, much of it aimed at Intel, and Intel in turn had grown willing to open its factories to outsiders in a way the old, dominant Intel would never have entertained. A company that needs work and a pair of companies that need a fabricator find each other naturally. The electricity a serious fab consumes is its own enormous problem, a leading plant drawing on the order of the power of a small city, and here the rest of the enterprise had something to offer that almost no other chip customer could, having spent a decade learning to generate and store and move power at scale. The fab would sit downstream of everything the energy layer had already been built to do.

The discipline Musk had drilled into Tesla and SpaceX had a name inside those companies: the algorithm. Question every requirement. Delete the part or process that does not earn its place. Simplify what remains. Speed up the cycle. Automate last. "If you're not adding things back in ten percent of the time," he had said at Starbase, "you're clearly not deleting enough." A fab would test that method against material that records every mistake. Change a process step, measure the yield, and learn whether the wafers carry more working circuits or fewer. The pace worth owning is the pace at which a line can find a fault and return with a sounder chip.

The project's name points to an ambition of manufacturing one terawatt of computing hardware a year. A terawatt is a trillion watts, roughly the output of a thousand large power stations; here the figure describes the combined power rating of hardware made in a year, rather than electricity generated by the factory. At that scale a chip design would have to become a flow of wafers, packaged devices, boards, power supplies and cooling systems, with every stage able to take the output of the last. A car that reads the road, a robot that handles a changing part and a cluster training their next models all depend on that flow. The proposed orbital computers would add another destination, with radiation and launch mass changing the package again. The first good wafer would be a small result beside the stated number, and the beginning of learning what a line could reliably make.

THE HISTORICAL ANALOGUE

Ford's own version of the move was treated, in its day, as a rich man's eccentricity. Observers thought it strange, an overreach by a man who could not stop acquiring. Within a decade it was the model that defined American industrial dominance, and the pattern repeated at every industrial era that followed. Carnegie had done it with steel, owning the ore and the rail and the mills. Rockefeller had done it with oil, owning the wells and the pipelines and the refineries. When the United States set out to win the space race a generation later, the federal government effectively subsidized the entire early microelectronics supply chain into existence around the Apollo program, buying the first integrated circuits in such volume and at such prices that an industry grew up to make them, because the chips the program needed had no commercial market yet to bring them into being.

The Apollo case is the one that rhymes most exactly with the present. When the integrated circuit was new and costly and unproven, almost no commercial buyer wanted it, and the young firms that knew how to make it had no market large enough to justify scaling up. The space program became that market. For a stretch in the early 1960s the Apollo guidance computer and the missile programs beside it bought integrated circuits in large quantities, giving manufacturers orders they could plan around, helping them build production lines and bring costs down, and turn a laboratory curiosity into the foundation of an industry. A government had looked at a technology too important to leave to the slow mercy of the market and simply willed the supply into being by becoming its customer. Sixty years on, the most advanced chip production rested heavily on the foundries of one island. A new set of builders wanted another place capable of doing that work.

In the same window, at Bell Labs, Mervin Kelly took the principle inside a single institution. Kelly, who ran the labs from the 1940s into the 1950s, treated it as part of his job to own the reliability and manufacture of the vacuum tubes the telephone network ran on. Western Electric, the Bell System's manufacturing arm, made the tubes; Bell Labs developed designs and processes for the network's demanding tolerances. The reasoning the Bell engineers gave for taking on all that trouble came down to consequence. A failed light bulb is cheap to replace and rarely urgent. A failed repeater out on a long line takes a great many conversations down with it, costs real money, and could in the wrong circumstance cost lives. System criticality drove manufacturing control. Vertical integration, in every one of these cases, was the form an industrial layer took once it had become critical enough that the operators above it wanted to set its pace with their own hands. If Terafab produced its first wafer, the rest of the stack would be waiting.

11

One Day, Four Sites

SET THE CLOCKS OF four working places to one October day. The machines are real; the day that brings their jobs together is imagined.

MIDNIGHT IN LATHROP

Eight minutes past midnight in Lathrop, California, and the night shift at Tesla's Megafactory is still assembling batteries for the grid. The cells inside these modules began farther upstream, held in formation racks for their first controlled charge and discharge. Their readings were checked there; a cell with an unusual result waited for review while qualified cells moved toward assembly. Here the work is enclosure, cooling, wiring, test, and the long business of making a unit that can leave the factory ready to stand outdoors for years.

Along the floor people in blue coveralls move between open cabinets, lifting, seating, scanning, and passing the work along under light that does not change with the hour. A finished Megapack waits at the far end for the freight equipment that will take it to a customer. The building sounds the way it sounded at noon. For the people on the line it is an ordinary shift.

The people working this shift have the flat unhurried competence of those who do the job while the rest of the valley sleeps. A woman at an open cabinet checks the cooling lines before the next assembly closes around them. A few stations down a man seats a module and scans its place in the build record, with the loose economy of motion that comes from having made the same reach thousands of times. Forklifts move along the marked lanes with their warning tones bouncing off the high steel ceiling. Outside the dock doors the Central Valley is dark, the yards and warehouses along the road showing their security lights. Inside, work keeps its own time. A station can meet its cycle and still leave completed assemblies waiting for a test bay; the queue says more about the night's output than the speed of any one pair of hands. At handoff the next shift needs to know what was meant to clear, what cleared, and what held the rest. The pass-down tells the next crew where its time will go.

At the far end of the floor, behind a rope of yellow tape and a border of stickers on the concrete, two Optimus units are carrying empty parts trays from a rack to a packaging skid. The pen they work in is about the size of a small office.

The task is deliberately small; nothing on the assembly line waits for the robots to finish it. One lifts a tray, turns, takes four steps to the skid, sets it down with the slight settling hesitation that the gait still carries, and turns back, while the other waits at the rack so they do not cross. An engineer in a hi-vis vest stands just outside the tape with a tablet, logging each cycle.

On her tablet the two machines show up as two lines on a chart, cycle time against the clock, and beside them runs a third line for a person doing the same handling task one station up the floor. The human is faster. When one unit lifts a tray that has not seated square on the rack it stops, holds it a moment, sets it back down, re-indexes its grip, and lifts again. The engineer marks the recovery, then watches the next pass. The yellow tape gives the test a boundary the line lead can see from across the floor. Inside it the robot has room to fail slowly and show the team exactly where the motion breaks.

The people on the floor have made their peace with the pen, or have stopped seeing it, which comes to the same thing on a night shift. The worker two stations up carries a finished part past it without slowing. Once an hour or so the line lead drifts over, reads the engineer's tablet over her shoulder, says something short, and returns to the floor. Most of the room still belongs to people. The robots work inside their square, repeating a task that can be measured against a person's hands, while the production line goes on around them.

Making batteries also takes power, much of it in loads that rise and fall as equipment starts, holds, and starts again. A row of storage cabinets beside a plant like this could take some of that swing. A dispatch system would watch the next demand on the line and the cabinets' state of charge, leaving enough in reserve for a shift change when several operations may start together. It could fill the cabinets during a lull, give power back when the draw rises, and keep the factory's connection inside what it can carry. Inside, workers would still fit the cooling lines and close the enclosures. The two robots at the end of the floor would keep making their slow trips inside the tape. The electrical schedule and the shift schedule would meet at a control point few workers needed to see.

THREE FORTY-SEVEN, THE HIGHWAY

At three forty-seven in the morning three Tesla Semis are east of Reno on Interstate 80, carrying crated control equipment and replacement modules from Lathrop toward utility storage sites farther east. The trucks left the factory earlier in the night and crossed the mountains while the assembly line kept running. In the lead cab a driver watches the road and keeps a hand close to the wheel. The highway is open in long stretches at this hour, but the on-ramps and grades still require his attention. The other two drivers are awake behind him, each responsible for a heavy truck and the load it carries. A line of red lamps moves through the dark beyond the last warehouses and into the basin country.

The lead truck's assistance system reads the lane and holds a gap under its driver's supervision. The road helps. Interstate 80 across northern Nevada runs wide and mostly straight through open country, well marked and lightly traveled at this hour, with the harder work at the on-ramps and on the long grades where a loaded truck has to manage its momentum. The driver watches the mirrors and the shoulder, ready for the part of the road the cameras or the plan cannot settle. The cab is quiet except for the hum of the drivetrain and the wind.

Later in the run the trucks will pull into a charging stop with stalls built for Semis. The cables are heavy enough that a driver takes hold of one with both hands; once connected, the power goes into the truck while its driver walks the lot, uses the restroom, and fills a coffee cup. Three trucks drawing hard at once would make the feeder, as well as the stalls, part of the route's capacity. Storage at a stop designed for this traffic could charge between arrivals and help carry the surge when the plugs go in. Charging belongs in the route plan just as surely as a grade or a delivery window does. The stop gives the three people a break and the trucks the energy to carry their loads on toward Utah. When the plugs come out, the drivers check their trailers and climb back in.

The load is equipment for energy storage, small enough in its crates to ride on ordinary trailers. The finished Megapacks taking shape in Lathrop are larger pieces of the same business, shipped on their own freight plan. Here the electric trucks move controls and replacement parts a utility crew will need at its site. At the charging stop a driver hangs the heavy plug on its stand, walks the length of his trailer, and tugs a strap where a crate has settled against its blocking. The truck takes its power from a connection beside the road, then carries the grid equipment on through the night.

Dispatch can follow the trucks' positions, charge states, and expected arrival times through the fleet system. The useful number changes with the work. A loaded truck climbing a grade spends energy differently from one running level; a longer stop at the charger moves the arrival estimate even if the road ahead is clear. A dispatcher looking at all three can see which load will reach its delivery window and which needs a call before the customer starts waiting. The driver needs a reachable charger and enough charge to make the next leg with the weight still behind him. Overhead, Starlink satellites cross the same dark sky. The trucks pass beneath them toward Utah; another pass can carry a Starship's telemetry after it leaves the Texas coast.

The lead driver calls the pace for the three trucks. Each following driver holds a safe interval, works the grades, and watches the traffic entering from the right. Software may take more of those small corrections over time; on this run the people still own the distance between the vehicles and the decision to keep moving. At four in the morning the lead truck's lamps tip over a rise. The second driver lets the gap open before bringing his own truck up after it, and the third follows through the dark.

SIX TWENTY-ONE, BOCA CHICA

The same morning, twenty-one minutes past six, on the Gulf coast at Boca Chica. A Starship sits on a pad undergoing pressurization for an afternoon static fire, venting thin and white into air that is already warm. Starbase built the vehicle in the high bays behind it and brought it out to the pad for this test. A row of energy-storage cabinets beside the ground equipment takes some of the electrical swings as pumps, valves, and instruments come on; inside the control building, ordinary racks of flight and ground-system electronics watch the vehicle. When a Starship flies, its telemetry can come down through the Starlink satellites passing over the coast.

The pressurization is routine and slow. Cryogenic propellant load comes later in the day; this is the gentler morning test, inert gas working up through the tanks to prove the vehicle holds its pressure, the venting that thin white plume the cameras along the beach pick up before the sun is fully over the water. Gulls work the tide line a few hundred yards off. The site is quieter than its reputation, most of the people who run it still arriving, the night's security handing off at the gate, a few trucks idling with their lights on. The vehicle on the pad is enormous and patient and doing almost nothing. Its gauges rise a little at a time while the ground equipment holds and measures the pressure; at Lathrop and on Interstate 80, different machines are also measuring changing loads and giving people time to respond.

Out on the pad the only motion is the venting, a thin steady plume working off the top of the vehicle and bending east on the morning breeze, and the only sound at this distance is the low hiss of it carried in over the flats. The heat is already building. By midday the steel will be too hot to lay a bare hand on, and the crews who have to work the upper levels are doing it now, in the cool, before the sun turns the vehicle into a tower of warm metal standing over the water. A tech on a man-lift rides up the side of it to check a fitting, small against the hull, and comes back down, and the vehicle goes on venting and holding its pressure and waiting for the afternoon, patient in the way only something very large and very full can be.

Boca Chica had grown into a small industrial city of its own over the decade, a sprawl of high bays, tank farms, and tracking antennas on what had been dune and tidal flat, with a cycle of building and testing that filled most days. The morning's pressurization was one line on a schedule that would fill this day. By afternoon the same vehicle would take on propellant, hold for the count, and light its engines against the hold-downs for a few seconds while the ground systems supplied the pad and the control racks logged every channel. Routine by the standard the site had set for itself. Loud enough to be heard across the river in Mexico.

Inside the control room the consoles were filling as the morning crew arrived, each operator taking a station and each station a slice of the vehicle, propulsion and structures and avionics and ground systems spread across the screens. At one console an operator read the rising tank pressure beside the valve state and the step in the procedure, then checked the trace through a hold before telling the test director that station was ready. The instruments watched thousands of channels and flagged what drifted; the people read those flags and judged what mattered. For a static fire the humans still held the final call on every gate. The test director at the console had a hand near the abort, and a vehicle under pressure waited outside for the room to agree it was ready.

Starbase makes the rockets that rise from here, while steel and propellant, food and spare parts arrive by road. Trucks keep the site supplied because even a place built to send vehicles into orbit depends on the ordinary freight lanes behind it. The battery cabinets in this day's pad arrangement are another way of feeding a short, uneven demand. The Starlink satellites above are the link a flying vehicle can use when it has left the coast behind. A technician checks a fitting on the ground system; inside the control room the test director watches the pressure continue to rise.

THE ARGUMENT SPACEX IS MAKING

At nine in the morning outside Austin, the Cortex 2 cluster finishes an inference run. The electrical draw drops, then climbs again as the next batch loads. Storage cabinets beside the site take part of that swing, charging into the trough and giving power back at the peak while the grid connection carries the rest. The cooling system and the operators have their own work to do. A rack of compute changes load much faster than a utility builds a substation, which makes the timing of power a design problem for every large cluster.

SpaceX proposed carrying computing into orbit. Computers in orbital shells would draw power from their own solar arrays and link to the network already above Earth. A cluster on the ground is still a large physical plant: processors take power in changing loads, cooling equipment carries away nearly all of it as heat, and a grid connection has to be built for both. The Cortex run ending outside Austin makes that demand visible in a small change on a meter. A new terrestrial hall needs transformers and a connection before its first useful computation. SpaceX is asking what part of that work could instead be made, launched and repeated as spacecraft.

A selected sun-synchronous orbit can keep a solar array in sunlight for much of its circuit. The array would feed processors that still have to be held inside a workable temperature range. Heat leaves a chip through its package and the spacecraft's thermal path, into heat pipes and out across radiator panels that emit infrared into space. The panels need area and a useful angle; every watt the chips consume eventually has to leave through that surface. A radiator too small for the load would set the pace just as surely as the cooling plant at Cortex. In orbit the electrical source and the heat sink take the shape of the flying machine itself.

The proposed computers would need a way to exchange their work with Earth. Optical links would connect the compute spacecraft; traffic would then pass through Starlink to authorized ground stations. A job and its results would cross a moving set of satellites before reaching a person on the ground. That route has a capacity and a delay of its own, so a long simulation or a batch of model training is a more natural early load than a decision needed at a street corner. The job could keep running through many orbits while the network brought back its measurements and, eventually, its result.

A factory would have to make these units in quantity, and a reusable launcher would have to carry them up often enough to replace failed hardware as well as add new capacity. Each unit would have to keep its solar array pointed, its processors fed and its radiator exposed while radiation and repeated changes from sun to shade worked on it. Google's small orbital-compute prototype reached orbit on a SpaceX flight and made contact. Its instruments would measure a processor's radiation exposure, temperature and physical stress, questions a commercial fleet would have to answer over years. SpaceX's proposed network is much larger than that test. Its practical size would be set by what factories, rockets, thermal hardware and ground links could keep working together.

If that sequence held, a team whose next model or simulation was waiting on terrestrial power could send the work to a machine supplied by sunlight. A weather group running many possible flood paths could use another long batch of computation while there was still time to decide where to warn people. The result would return through the mesh to the people reading it. The immediate work is smaller: make one processor survive, move its heat to a radiator, hold a link through the moving network, and learn how often a working unit can be sent up to join it.

THE PIECES AND THE PATTERN

Four sites, separated by miles and working on different clocks. In Lathrop, a cabinet waiting at test can hold the line's output even while the preceding station keeps its cycle. On the interstate, the freight schedule moves with the truck's charge and its load. At the pad, a number on a screen matters in the step that produced it. At Cortex, a batch ends and the electrical load falls faster than a new substation can be built. Each place has a different way of finding the next constraint and a crew responsible for clearing it. Once a test bay opens, a charger becomes available, or the control room clears a gate, work that has been waiting can move again.

Each site has its own energy problem. Upstream, cells take their first controlled cycle before they can be assembled for grid storage. On the highway, the truck's charge must cover its load, the grades, and the distance to the next stop. At the pad and the compute cluster, equipment draws power in steps the local connection has to carry. Material takes its slower road in assemblies, crates, and the steel carried from a high bay to a pad. Compute stays close to the task it serves, reading a lane, correcting a grip, sampling a vehicle, or taking the next batch. The state of each job travels to someone who can act on it, in a build record, a dispatch screen, a test log.

THE PAD AT LAST LIGHT

Late afternoon at the pad. The static fire is done and the engines have gone quiet, and the vehicle stands in the heat shimmer venting a last thin thread of vapor into the Gulf sky. Beside the ground equipment the gray storage cabinets sit in a low rank, their cooling fans still turning. Behind the blast wall, in a windowless room, the control racks that watched the burn keep running, and a single test engineer sits at a console reading the numbers down off the screen, the only person in the frame, while above the whole arrangement the satellites that can carry the next flight's telemetry come around again on their pass over the Gulf.

12

Where the Layers Meet

A CAR'S ELECTRICAL SPINE

AT A WINTER TEST ground in Sweden in early 2026, reference vehicles from Volkswagen, Audi and Scout carried a new electrical architecture through cold-weather trials. The joint engineering team had put powerful central computers and zonal controllers into vehicles built by different brands. A command for a light, a window, a display or a motor still had to reach the right part at the right moment, even as the temperature changed what batteries, connectors and materials would tolerate. The tests were one step between a working prototype and a car a factory could make in volume. A failure here would travel back through both the software and the wiring before either was fixed for a production line.

The work had begun in warmer air. The teams first prepared functions and reference vehicles in Arizona, then took the systems through snow and ice in Sweden. All-wheel drive, traction control and driving response made the relation between code and hardware observable. The wheels had to receive the commanded torque at the surface they were actually touching. Engineers also tried remote software updates, since the promised ability to change a vehicle after sale is useful only when a new version can be installed and checked without disabling something the owner already relies on. The teams put the design through hundreds of test and validation cycles. The result was a production-intent architecture through one demanding stage of testing, with more development still ahead before a customer vehicle left the line.

The architecture came from Rivian and Volkswagen Group Technologies, a joint venture founded in 2024. Volkswagen had put a substantial commitment behind it, up to $5.8 billion in Rivian and the venture, and Rivian had brought an electrical design already running in its own vehicles. The two companies built a drivable Volkswagen demonstrator in twelve weeks. By the following winter a larger team was testing the design in reference vehicles for several brands. The point of the arrangement was visible in the hardware. A zonal design gathers functions under a smaller number of powerful computers, giving engineers a way to change vehicle behavior through software without opening up a separate controller for every feature. What works in one model may then travel to another, provided the physical interfaces and the tests travel with it.

There is another working line behind those trials. At Wolfsburg, the plant extends toward the horizon under the four chimneys of its power station. Body parts meet at fixtures; welds, paint and final assembly have to arrive at one pace; a car is inspected before it leaves the line. Generations of workers and engineers have learned how to keep that pace through model changes and interrupted supplies. A controller designed far from the plant has to join that discipline. Its connectors must fit the harness, its thermal behavior must survive a closed body in summer, its software must still speak to a replacement part years after the first car was sold. The factory's skill is part of the new architecture, because the architecture has to be made and serviced as a car.

The joint venture joins companies that can each do work the other needs. Volkswagen brings a vast range of vehicles, manufacturing and service demands; Rivian brings a design formed around central computing from the start. The boundary between them does not vanish when they agree to work together. A change to a controller has to reach teams responsible for software, electrical hardware, vehicle safety, purchasing, production and repair. One partner may see a fault in a test car while another owns the code or part that can remove it. Shared specifications, access to the same failure data and authority to change a design are the practical substance of the venture. A contract creates the place to meet; the engineers have to make the interface hold.

Volkswagen's China teams took another route with Xpeng and a locally developed zonal architecture. Production of that electrical design began in January 2026, and the first jointly developed car came off a line that March. Different roads and customers ask for different vehicles, and a global carmaker cannot bring every model onto one electronic design overnight. The programs had a common burden. A century of skill at making the visible automobile had to join skill at making the code and the electrical spine that now carry so many of its functions. In one region a joint design was reaching a car built for sale; in another a reference vehicle was returning from cold-weather tests. The same manufacturer was learning two ways to make an electronically different car, then had to make each of them serviceable for years on its own roads.

That knowledge has to move between people as well as parts. Volkswagen announced a program to send its software specialists into the joint venture's workplaces, including Palo Alto, for months of direct work with the architecture and its code. On returning to Wolfsburg, they were to bring what they learned into the brand's development departments; Audi and Porsche were planning similar training. The plan recognizes a practical limit of buying a design. A local engineer must know enough to tell whether a new model's behavior came from a sensor, a network delay, a line of code or the physical car. The joint venture can build a common spine, but each brand has to learn how to give that spine a body.

A CAPSULE AND ITS RETURN

In June 2024 Boeing's Starliner carried Butch Wilmore and Sunita Williams to the International Space Station on its first crewed flight. During the approach, reaction-control thrusters underperformed and helium leaked from the propulsion system. Engineers on the ground and in orbit worked through the data. They fired jets while the capsule was docked, compared thrust and pressure readings, tested a thruster at White Sands and studied the helium loss. NASA judged that the uncertainty was too large for a crewed return. Starliner came home without its astronauts. Wilmore and Williams flew back the following March aboard a SpaceX Dragon after months of work on the station.

The flight was a test, and the test returned knowledge as well as a spacecraft. The service module that held the troublesome hardware would be discarded before landing, so the team had to take as much evidence as it could while the vehicle remained docked. A thruster's response under a short firing, the temperature that followed it and the rate of a helium leak became pieces of the next design decision. NASA's choice protected the crew while Boeing and the agency continued the investigation. Their engineers still had the demanding work of finding the causes, changing the hardware or its operating rules, and proving the change on a later flight before ordinary crew service could begin.

NASA had hired Boeing and SpaceX in 2014 to develop separate ways to carry crews to the station. Dragon began crewed flights in 2020 and by the time of Starliner's test had made the trip repeatedly. Its Falcon 9 launcher could return a first stage to be inspected, repaired and flown again. Starliner's ride to orbit was a different launch system, and the trouble NASA was investigating was in the capsule's propulsion system. One crew system had accumulated many flights and recovery inspections while the other was still turning its first crewed flight into the data needed for certification. Each additional flight gave a team another chance to learn what wore, drifted or behaved differently from the ground test.

A crew vehicle is a chain of interfaces, from the part supplier's valve and thruster to the ground test stand, flight software, mission rules and the person who signs the return decision. Some of those teams sit inside one company and some across contracts. That boundary can be made workable when the people who find a fault can reach the people who made the part, run a test that reproduces the failure and carry the result into the next vehicle. The same need appears in a car's zonal wiring. A fault report travels through a human organization before it becomes a different wire, line of code or piece of metal. The distance it travels is a design problem in its own right.

The two commercial crew vehicles gave NASA another kind of capacity. When Starliner could not carry its first crew home, the agency could assign them seats on another system that had already been flying. That redundancy had taken years of separate engineering and testing to build. It is a reason to keep unlike approaches alive, even when one reaches routine service first. A second route becomes valuable at the moment the first route needs time to learn from its failure.

THE PRODUCTIVE SEAM

Some boundaries have made whole industries possible. In 1981 IBM assembled its personal computer with an Intel processor, a Microsoft operating system and other parts it could obtain without making every one of them. The arrangement brought a machine to market quickly. IBM published enough of its design for other makers to build compatible hardware, and the computer became a platform for software and equipment IBM had not conceived. Intel could improve a processor for many computer makers; Microsoft could write an operating system that ran on their machines. The very openness that let the PC spread also gave other companies room to make machines that competed with IBM's. The boundary was a source of reach and a source of lost control, both consequences of the same architecture.

Morris Chang founded TSMC in 1987 to manufacture chips designed elsewhere. To make that split workable, the foundry had to give each designer a usable account of its process before anyone paid for a mask.

TSMC's design kits turn part of that account into a tool another company can use. They specify shapes a process can print, electrical models for what a transistor is likely to do, and standard blocks that can be assembled into a larger circuit. A designer tests an idea against those constraints before committing to a costly mask. The foundry runs test vehicles of its own, reads defect patterns and changes recipes until the yield is reliable enough for customers to risk their designs. When a failure appears, a useful report says whether it followed a particular layout, a material, a tool or a step in the process. That is coordination across a corporate boundary at the scale of a few nanometers. It has allowed a small team with an original circuit to start from a manufacturing capability built by thousands of other people.

A car asks for a different arrangement at some of its seams. A battery maker can supply cells to several vehicle companies, each with different packaging and cooling. That specialization may improve the cell faster than a single carmaker's captive line. But once cells enter a pack, the enclosure, heat path, charging controls and crash structure must behave as one object. A change in chemistry may ask for another cooling channel or a new charging rule. Keeping those people in one company can shorten the loop; a supplier and vehicle maker can also shorten it if they share test data and authority across the boundary. The decisive question is where feedback slows down, and whether the chosen organization can carry a fault back to the people able to remove it.

Europe's automakers grew out of a dense field of such specialists. Bosch and Continental made components and systems that crossed brand lines; a car plant drew on a network of tooling, materials and electronics far beyond its own gates. That network gave the continent factories of extraordinary quality. Software changed the load on the interfaces. Code that updates after delivery needs to know which controller, sensor and actuator sits in each vehicle; a repair in one layer may expose a failure in another. The public Volkswagen-Rivian and Volkswagen-Xpeng programs show companies building new ways to carry that work. Their results will be visible in production vehicles whose functions can be changed, diagnosed and repaired over time, with supplier and carmaker sharing enough of the fault record to keep the system sound.

The old modular network still has jobs no one car company should try to swallow. A supplier making a brake sensor for several manufacturers can see how it fails across different vehicles and climates. A foundry serving many chip designers can keep tools running at a volume none of those designers could support alone. A charging connector standard lets a driver use a network built by people who never designed the car. The driver stops at an unfamiliar station, fits the plug and gains another stretch of road. That small compatibility gives a car a range larger than the network its maker could build alone.

FROM THE CELL OUTWARD

BYD began as a battery company in 1995 and entered the automobile business in 2003. Its first expertise was in a small, repeated electrochemical object; the vehicle gave that object a larger use. Cells had to become packs, packs had to fit into bodies, and a body had to protect both the people inside it and the material storing its energy. Each step brought another manufacturing discipline close to the battery line. At enough volume, a change in the cell could change the way a vehicle was packaged, while a problem in the vehicle could return as a request to the cell makers. The short path for that physical lesson helped the battery and the car develop together.

Its cell-to-body design makes the connection unusually literal. The Blade Battery is set into the vehicle structure rather than treated as a sealed box carried beneath an otherwise finished car. That arrangement asks the cell, pack housing and body engineers to agree about stiffness, crash loads, heat and service access. A cell whose dimensions change for better energy density can alter the space available for a floor or cooling passage. A body made lighter by using the pack as structure can change how many cells the vehicle needs. The engineers can trade those properties against one another while the product is still being designed, then test the finished car to see which gains survived manufacture. Integration gives them more variables to adjust together and more consequences to check together.

CATL worked the battery layer on a different scale. It supplied cells and storage systems to customers with unlike vehicles and grid sites. A customer needing a safer pack or longer life can give the manufacturer a new test; a change in the manufacturing process can then serve several customers. In a cell plant, coating thickness, drying and formation are repeated on an immense number of units, and small shifts in yield matter across a whole year's output. CATL's position as a specialist lets that learning spread. The companies that build the surrounding car or grid cabinet still have to make the pack's electrical and thermal connections sound. Specialization can increase the number of builders able to begin from a good cell.

That manufacturer can also change the chemistry on offer. CATL's sodium-ion storage work, for instance, joins materials research to the less glamorous business of getting a large number of cells to behave alike. Its hard-carbon process called for moisture control and changes along the line aimed at more consistent output. A grid builder can use an alternative chemistry only after it knows what heat, lifetime, controls and safety margins the installed cabinets require. The handoff runs in both directions: the builder specifies the duty cycle, while the cell line determines what can be delivered repeatedly. More chemistries give builders more ways to fit storage to a place, but each new one has to pass through that production and installation chain.

China's industrial system also builds large connections beyond a cell maker's own reach. Transmission lines carry power across provinces; rail and ports move materials between plants and markets; public investment can make a site reachable before a single firm can justify the whole route. Their capacity meets the battery factory's at substations, loading bays, standards and purchase orders. A public line that carries dependable power can give several independent workshops a place to operate. The workshops still have to raise yield, keep trained people close to difficult processes and turn a usable part into a product someone needs.

There are limits the most coordinated purchaser cannot erase. Leading-edge chip production needs tools, process recipes and experienced teams accumulated over many generations of manufacture. China has invested in domestic capability while export controls restrict access to some of the newest lithography equipment. Designers elsewhere depend on foundries and equipment makers too. A shortage at one layer can hold up work in many places, which is why foundry capacity, toolmaking and the education of process engineers are more than corporate assets. They determine how many different builders can put a new circuit into a working machine. The constraint can be approached through more places to make and qualify chips, more compatible designs and better use of the silicon already available; it cannot be wished away by drawing a taller organization chart.

The car, the battery and the grid each accumulate knowledge on a different clock. A model year may change a body; a cell chemistry takes years to qualify; a transmission line may serve generations of both. One organization can bring some of these clocks together. Another can keep the interfaces stable enough for separate specialists to work at their own pace. BYD's cell-to-car path can give a vehicle a body shaped around its battery. CATL's cell-to-many-customers path can put storage into several kinds of vehicle and into a grid cabinet that carries power after the sun is gone.

WHAT A BOUNDARY HAS TO CARRY

Ford's River Rouge offered an earlier answer to the problem of distance between stages. Ore, glass, steel and rubber entered a vast complex; vehicle parts and then completed cars left it. Tools and workers could be arranged for a regular flow, with less waiting between one shop and the next. Ford could see how a change in material affected a later operation and could set a tempo across the whole plant. The physical achievement was immense. When the Model A replaced the Model T, however, a new assembly line at the Rouge took time to reach full output. A system built to move one design smoothly must be reworked when the design changes. Tight coupling buys a fast path for one kind of feedback and makes a large change travel through many machines.

The change had a human clock. The Model A was presented in late 1927, yet its new Rouge assembly line did not reach full capacity until the middle of 1929. Toolmakers had to prepare a different car, workers learn its operations, suppliers feed different parts, and the early line had to reveal which steps failed at the new tempo. The plant was still an extraordinary way to turn raw material into a vehicle. Its slow climb into the new model showed the cost of asking one tightly joined system to turn all at once. Later builders could use that knowledge when deciding which machines to dedicate to a particular shape and which interfaces to leave open for a new one.

The dedicated chip foundry answers a different part of that problem. It holds the expensive, exacting process steady enough that many designers can change their circuits without each rebuilding the factory. IBM's open PC architecture gave other firms a common box into which they could put their own software and peripherals. The vehicle partnership at Volkswagen tries to hold a shared electrical design steady across several brands while the brands keep making different cars. Ford's line asked one plant to turn with a model. The foundry lets a hundred layouts change while its process is kept within tolerance; Volkswagen is testing whether one electronic spine can remain reliable beneath cars with different bodies and uses.

The choice becomes clearer at a failed part. If a thruster loses performance, its test trace must reach the people who can change the valve, the thermal environment or the rule for using it. If a wafer loses yield, the designer and foundry need to locate whether the trouble came from layout, mask, material or process. If a winter-tested car loses a function, its software team, hardware designers and vehicle engineers need to reproduce the fault in a form the production crew can check. A company can place all those people under one roof and still fail to pass along a useful measurement. Several companies can pass it well if their interface includes the evidence, a shared test and a route to a decision. Organization matters through the work it permits after something goes wrong.

Those choices put other work within reach. A design team with a circuit for a medical instrument can use a foundry's qualified process without first financing a fab. A grid operator can buy a tested cabinet and hold afternoon power for the evening load. When Starliner returned empty, Dragon gave Wilmore and Williams the working route home. These uses depend on many trials and on people able to read a bad result. Once the work is repeatable, the instrument, the cabinet or the flight can become part of someone else's ordinary plan.

WOLFSBURG AT THE TURN

The Wolfsburg plant still runs toward the horizon under its four chimneys, with a city's worth of skills gathered around the repeated act of making a car. Years of solved interfaces sit inside every finished vehicle. A stamped panel meets another at a fixture; a weld must be sound; a cable must reach a connector without rubbing through its insulation; the test at the end of the line must catch what a customer should never have to discover. Those are things trained hands and instruments have learned to do together. The arrival of central computers and zonal wiring gives them another set of interfaces to master.

The cold-weather reference vehicles have come back from test. Their measurements have to reach designers who can use them, and a revised part has to survive another trial and, eventually, the rate of a production line. Volkswagen's engineers can draw on the plant's long memory of making that passage; Rivian's engineers bring an electrical architecture built on another line. The planned months of shared work would return software specialists to Wolfsburg with a closer knowledge of both the code and the people able to change it. What emerges will depend on whether these forms of knowledge can meet at a specific failure and change the next vehicle. At Wolfsburg the line keeps moving. The test records from the cold are back with the people preparing another car.

PART IV

THE CIVILIZATION STACK

An Austin street, an orbital yard and a ship bound for Mars begin with machinery already taking shape. Their days are imagined; the work that might bring them nearer is under way.

13

Earth, Re-stacked

MORNING, AUSTIN

ON A RESIDENTIAL street in Austin, on a clear morning late in the decade, a Cybercab pulled to the curb at forty-two minutes past seven to carry someone to the airport. It had no driver and no steering wheel, and it arrived at the exact minute it had said it would. As it eased back into the lane it slowed for a moment, because an Optimus was crossing the street ahead of it with a tall pile of grocery bags balanced against its chest, walking the delivery up to a porch four houses down. The car waited for the robot to clear the lane and then went on. Two streets to the east, at a substation behind a low wall, a row of Megapack units was carrying the morning surge as the neighborhood woke and switched on.

On the roof of the house the Cybercab had just left, a small Starlink terminal angled at the sky was carrying a video call in the kitchen, a grandmother in another time zone talking to a child eating cereal.

Inside the house, the morning ran on the same quiet machinery as the street. The airport trip had been booked the night before with a sentence spoken into the kitchen, and the car had arranged itself around the household's morning without anyone confirming it twice. The coffee maker had started on its own at half past six. The thermostat had read the day's forecast off the network and cooled the rooms ahead of the heat that was coming. A parent moved through the kitchen packing a bag, half-listening to the grandmother on the screen, glancing at the time on the wall, and at no point in the morning did anyone in the house think about where the electricity came from, or how the car knew when to arrive, or what held the call open across an ocean. The morning simply worked, and a working morning asks nothing of the people inside it.

Up the block a second Cybercab sat at another curb with its door open, waiting on a man who was still inside finding his keys. It waited without complaint and without a meter, because there was no meter and no driver standing idle to be paid for the minutes. A woman three houses down still drove her own car, an older Tesla she kept out of habit and a fondness for the wheel, and she backed it out of the driveway by hand while the autonomous traffic on the cross street slowed, opened a gap, and closed it after she entered.

The horns stayed quiet, because there was nothing to be impatient at and no one driving them to feel it.

The street was quieter than a street like it would have been ten years before. The cars ran on electric motors, and the loudest things in the morning were birdsong and the hiss of a sprinkler two yards over. A jogger went past with earbuds in, talking to someone who was not there, her words handed off from the phone on her arm to a tower or to the low satellites overhead, a routing she would never see and never wonder about. A delivery drone crossed high over the rooftops on its way somewhere, and no one looked up. The weekly garbage truck had learned the street so completely that it no longer hunted for the bins. It swung its arm to the exact spot where each one stood and set it back down square with the curb.

Down the block the Optimus that had delivered the groceries had a dozen more stops to make before the morning was out, and it knew the neighborhood in route-level detail, including which gate stuck, which dog to give a wide berth, and which porch step had come loose. Two children went past it on their way to school, and one said hello to it by a name the family had given it, and the robot said good morning back and stepped aside to let them by. It had learned the route by walking it morning after morning, never once told where to go.

Inside the first Cybercab, the passenger on the way to the airport answered email and drank coffee from a paper cup. There was no road to watch in the old sense, only a windshield full of an ordinary morning sliding by, while the car took the on-ramp and folded itself into a stream of other driverless vehicles running close together at a steady speed, each one leaving the gap the cars behind it needed and no more. The fare had been quoted before the trip began and settled without a card or a tap, drawn from an account the rider had set up once and long since forgotten. The car would reach the terminal at the minute it had named when it pulled from the curb, because it had already counted the lights and the merges and the morning's traffic into the promise.

On a porch across the street an old man sat with his coffee and watched the morning. He had driven trucks for a living once, long hauls across three states, sleeping in the cab and fighting the weather and the clock, and he watched the driverless Semis run those same roads now with something between admiration and disbelief. He had been there for the entire change and could still feel the strangeness of a loaded rig rolling through the dark with no one in the seat. To the children walking past him to school it was no stranger than rain.

THE SUPPLY CHAIN YOU DON'T SEE

Underneath the ordinary morning was a machine almost no one in the neighborhood ever looked at. The grocery bags the Optimus was carrying had arrived in the city overnight, on an autonomous Tesla Semi that drove in from a distribution warehouse ninety miles out while the street slept. That

Semi had been dispatched by no one the household could have named. A scheduling system somewhere had read the orders coming in from a thousand porches, bundled them into a route, matched the route to a truck with charge enough to run it, and timed the truck's arrival at the warehouse dock to the minute the loading robots would be free to fill it. The cab at the front of the Semi was a formality the regulations still required and the truck no longer used. The seat rode empty, kept for a law that had not yet caught up to the road.

The warehouse ran on Megapack-buffered solar, charging its arrays through the afternoon and drawing them down through the night, so that the building never asked the grid for anything it had not already stored. The energy layer underneath that warehouse was the same Tesla energy-and-charging operation Mike Snyder had run for years, now reaching past the cars and the home batteries and into the freight backbone, buffering the warehouses and the depots the way it had once only buffered driveways. The vegetables in the bags had been grown in a climate-controlled facility a few hundred miles south whose lighting and air and water ran on the grade of compute and the grade of inference that drove the Cybercab. The water that fed those rows moved through the building in a closed loop that lost very little to the air, metered by sensors that decided plant by plant how much each one needed and when. The light came off the facility's own roof and its own bank of Megapacks, and on the rare cloudy week when the roof fell short, the building bought power from the grid in the hours it was cheapest and held it against the hours it was dear. Every one of those choices was made by software running on the family of chips that drove the car, and none of them waited on a person to approve it. The signal that had coordinated all of it, the warehouse to the truck to the routing to the doorstep, had moved through ground-station-to-orbit-to-ground links the household would never see and never need to think about.

To do any of this the machine had to know where everything was, and it did, in detail no single human mind could hold. Every curb the Cybercab pulled to, every porch the Optimus climbed, every lane closure and pothole and repainted line in the metro lived in a map the machines kept current among themselves, each vehicle reporting what it saw and reading what the others had seen. The map was never finished and never wrong for long. A crew that closed a lane at dawn found the traffic already routing around it before the first cone was fully placed, because a car that had passed a minute earlier had seen the crew and told the rest.

The breakfast on the kitchen table had come the same invisible way. The coffee had been roasted in a facility that timed its batches to the price of power. The cereal had ridden a Semi like the one that brought the groceries. The milk had come from a dairy a hundred miles out where the cows were milked by machines and the herd was watched by cameras that flagged a sick animal days before a farmer would have caught it by eye. The paper cup the airport passenger drank from had been pressed from pulp at a mill that ran its dryers on the cheap night power and idled them when the grid asked. Nothing on the table announced where it came from, and no one at the table asked.

All of this leaned in the end on cheap and abundant inference, the capacity to make millions of small judgments a second and act on them. The car reading a cyclist's intention at an intersection, the warehouse deciding which order to pick first, the grid guessing the evening's demand off the weather and the calendar, the greenhouse choosing how much water a single plant should get, all of it was the same act, a guess made well enough and fast enough to be trusted without a person checking it. A decade earlier that kind of judgment had been scarce and costly and reserved for the few places that could afford it. By the late 2020s it had grown as cheap as electricity and ran underneath the day in that unnoticed way.

The Cybercab carrying the airport passenger made a thousand small decisions a minute and showed none of them. It read a ball rolling into the street two blocks ahead and eased off a moment before a child appeared chasing it. It picked a lane for a merge that was still half a mile away. It let a delivery van cut in because the van's own system had asked, the two machines settling the courtesy between them in a tenth of a second. The passenger felt only the smooth forward motion and the coffee steady in the cup, and read email all the way to the gate.

The morning the household woke into was the quiet end of a night the machine had spent working. While the street slept, the busiest hours of the system had come and gone. Trucks had moved between cities on empty highways. Warehouses had loaded and unloaded under lights that dimmed themselves when no robot needed them. Data centers a few counties over had drawn their deepest power of the day to train and retrain the models that ran everything else, and the orbital links overhead had carried the coordination from one dark time zone to the next without pause. By the time the first Cybercab pulled to the curb at forty-two minutes past seven, the hard part of the day was already behind it, done in the dark by machines that did not need the light.

THE FACTORY HAS MOVED

Somewhere along the way the word *factory* had come loose from the thing it used to mean. A factory had once been a single site that made a single kind of object, a building you could point to. The car that had just pulled away from the curb was assembled in Texas from castings poured in Mexico, out of aluminum that had come from a smelter in West Texas, and the smelter ran on grid power buffered by a Megapack farm half a state away. No single building contained the process. The network was the factory, and the car was what the network produced.

The casting plant in Mexico ran its presses against the same demand signal the assembly line in Texas read, and both of them watched the price of power and the state of the roads and adjusted without a phone call between them. When a freeze shut a highway one winter morning, the network rerouted a week's worth of castings through a southern corridor before most of the people in the plants had finished their coffee. The decision touched four states and two countries, and no one in particular had made it.

There had been a time when a factory like the one that made cars was the center of a city, when a single plant employed many thousands of people and the town around it lived and died on the one company's shifts. That arrangement still existed in places, and some of the largest plants on Earth still ran that way. But the new way of making spread the work across a continent and across a thousand small sites, most of them modest, few of them the sort of cathedral a city grew up around. A casting shop here, a battery line there, a micro-factory printing parts on the edge of a metro, a greenhouse, a smelter, all of them stitched together by signal into something that produced more than any single plant ever had, and spread its people in ones and tens across a map where a town had once concentrated its workforce. A shop of ten could fill an order for a car line a thousand miles away and use the same machines for designs of its own.

On one of those benches lay a drawing no carmaker had sent. The shop's workers wanted to build a powered chair that could take a person across the broken curb cuts and sloping pavements around them. The vehicle brackets paid for the machines; the chair was the reason they stayed after the brackets were boxed. They cut a first frame, fitted standard cells and motors, and pushed it over the bad crossings to learn where its wheels lifted and where the motor ran short of torque. A bracket cracked at a joint that looked sound on the drawing. They brought it back to the mill and changed the radius. A replacement bearing arrived with the next order of vehicle parts. They fitted it, carried the revised frame to the curb and pushed its front wheels against the slope that had thrown the first version sideways. There were still crossings farther along the street they had not tried.

When a Cybercab wore through a part, that part was made on demand at a regional shop close to the cars it served, stamped or printed the night before it was needed and installed while the car charged, so that a vehicle could keep itself running for years on a maintenance schedule no owner ever saw or set. The fleet watched its own wear, ordered against it, and healed itself a part at a time. A car had become a thing the network kept making, in small corrections, for as long as it stayed on the road.

The materials moved in loops as much as in lines. The aluminum in the casting had been something else before it was a car, a window frame or a drink can or the body of an older vehicle that had reached the end of its road, melted down and poured again, because remelting aluminum costs a small fraction of the power it takes to pull it new from ore, and power was the thing the whole system counted. The battery in the Cybercab carried metals that had been through the loop before and would go through it again, pulled out at the end of each life and fed back to the front of the line. The network made things and shipped them outward, and it also took them back and made them again, until the line between a finished product and the raw material of the next one had worn thin.

The casting was poured a thousand miles from the assembly line and arrived exactly when the line was ready for it, because to the network a thousand miles was a number to plan around, an interval the system had already accounted for before the line ever called for the part. A piece could be made on one side of a continent and fitted on the other inside a day, and the household that drove the finished car would never guess how far its parts had traveled to find each other.

The smelter that poured the aluminum was itself a creature of the grid. Smelting metal is among the hungriest things a society does with electricity, and the West Texas plant had been built where it was because the wind and the sun and the Megapack farms kept power cheap there for long stretches of the year. When the wind dropped, the smelter could throttle its draw down within seconds and let the grid breathe, and it earned money for the throttling, so that the furnace had become a battery in its own right, a buyer of power that could turn itself down on command. The metal in the car had been shaped as much by the price of an electron as by the heat of a flame. Geography had become a variable the network optimized against, a set of distances and energy prices and latencies the system solved fresh every day, moving a casting here and an electron there to keep all of it humming at the lowest cost it could find.

THE GRID BREATHES DIFFERENTLY

The electricity grid the neighborhood drew on had learned to breathe. By the late 2020s it ran as a managed mesh, with power moving both ways across it. Houses generated power in the afternoon and gave it back. Megapack farms inhaled the midday excess and exhaled it into the evening peak. The Cybercabs charged themselves in the small hours when power was cheap and fed a little of it back when the grid strained at six in the evening, so that several million car batteries had become, without anyone deciding it as a single act, a layer of storage the grid leaned on.

It had happened the way large changes usually happen, without a launch date. It had been built up one rooftop and one car and one Megapack farm at a time, each addition too small to notice on its own, until one summer the regional operator looked at its own numbers and saw that the storage hanging off the edges of the grid had grown larger than the biggest power plant on its grid. The cars alone, parked and plugged in across the metro through the working day, held more energy between them than the region drew at its evening peak. The grid had grown a reservoir without anyone digging a lake.

The house on the residential street was a small piece of the same mesh. Its roof made power through the day, its wall battery held what the house did not use, and the Cybercab in the driveway, on the evenings it stayed home, could pour its charge back through the meter if the grid asked. The previous winter a storm had knocked out a transmission line for most of a night, and the street had stayed lit. Each house had run on its own stored sun until the line came back, the porch lights and the refrigerators and the routers carrying on while the wider grid went dark beyond the neighborhood, and in the morning a few people mentioned it and most had slept through it.

The roads had learned the same trick as the grid. The cars on them talked to one another and to the lights, and the morning rush had stopped being the standing river of brake lights it had been for a hundred years. Traffic moved in loose coordinated waves, the vehicles spacing themselves and timing the intersections among themselves, so that a crossing that had once needed a signal cycling on a fixed clock now let the cars thread through it in the order that wasted the least time. Accidents had grown rare enough that a fender bender drew a small crowd. The road, like the grid, had become a system that ran itself and asked for a person only when something at its edge went wrong.

What the grid had done, the other systems were doing too. The water that ran to the houses was metered and modeled the same way, leaks found by the pattern of the flow before a drop showed on a ceiling, pressure shifted through the day to follow demand. The cellular network handed calls between towers and satellites without a syllable lost. Even the money moved like this now, payments clearing in the background of a tap, the bank a piece of software the household touched once a month and otherwise forgot.

Multiply the one street by a city and the city by a country, and the same picture held at every scale. Ten thousand neighborhoods ran the same quiet way at once, each of them a few houses and a few cars and a substation, all of them stitched into the same mesh, none of them aware of the others, all of it balancing itself across a continent before breakfast.

In the control room that watched over the regional mesh, software forecast the day's load and proposed where stored power should go. During a heat wave a feeder began running hotter than its usual trace. The operator on shift checked the forecast against the live measurements, held some reserve for the evening, and called a field crew to inspect the line while the batteries carried a share of the neighborhood's load. The crew found a connection that needed work. The operator could give them a repair window before the feeder failed, then watch the temperature settle when they returned it to service. The machinery gave her time to act while the street kept its lights.

THE MUNDANE MIRACLE

Each piece of this would have been a product launch a generation earlier, a thing with a name and a keynote and a price. The robot that carried the groceries, the car with no wheel, the battery that ran a house through a storm, the phone that worked from anywhere on the planet, each had arrived as a marvel and been written about as one. And each had then done the thing marvels do when they keep working, which was to stop being marvels. They slid below notice one at a time, the way the dial telephone, the home refrigerator, and the elevator had slid below notice, until a person could pass a whole morning surrounded by them and remark on none.

There had been mornings, not many years before, when it did not work. The early version of the grid had failed under heat and failed under cold, and a whole region had once gone dark for days when a hard freeze knocked the old plants offline faster than the grid could cover. People who had lived through those mornings still felt a small private relief every time the lights stayed on through an August afternoon. The children eating cereal felt nothing of the kind, because to them the lights had always stayed on. The reliability an older person registered as a mercy registered with a six-year-old as the plain shape of the world.

Underneath all of it the numbers had been falling for years. A mile of travel, a kilowatt-hour of power, a delivered meal, a thing made and shipped, each cost less in real terms than it had a decade before, and the curve had not flattened. On the street it showed up as the small daily fact that the airport ride cost a few dollars, that the power bill had stopped being a thing anyone watched, that the groceries arrived for less than it would once have cost to drive out and fetch them. The savings came as the steady, unremarked cheapening of the floor everyone walked on, a little less every year, never enough in any single month to mention.

THE CHILD AT THE WINDOW

The household's six-year-old, the one who had been eating cereal through the call, climbed onto the back of the couch and looked out the front window as the Optimus walked back past the house with its arms empty now. The child watched it for three seconds, the way you watch a passing dog or a delivery van, and then turned away to find a shoe. Through the glass the street kept moving in its small silent motions, a Cybercab turning at the corner, another easing up to a curb across the way, a sprinkler throwing its slow arc over a lawn three doors down, and the morning rearranged itself the way it did every day while almost no one watched. The child had never lived in a world where this was not ordinary.

14

Orbit as Industry

THE VANDENBERG LAUNCH

A FALCON 9 LIFTED off from Vandenberg on a gray morning on the California coast, climbed through the marine layer, and was out of sight in under a minute. Inside the fairing was a batch of the second-generation Starlink satellites, each one larger than the first generation it was replacing, each carrying optical links that let it talk to its neighbors by laser, each built to work about five years on orbit before it burned up and a fresh one took its place. The booster came down on the drone ship offshore a few minutes later, and the upper stage carried the satellites the rest of the way to their deployment orbit. By evening, the launch had already sunk below the threshold of news. The cadence had become weather.

By that morning Vandenberg ran like a working port. The pads turned around in days. Propellant trucks came and went on a schedule posted weeks in advance, the range crews cleared one launch and began counting toward the next, and the recovered boosters rode back inland on flatbeds to be inspected and set up to fly again. A launch from the California coast had stopped drawing a crowd. The people who worked the range handled a Falcon 9 morning the way longshoremen handle a container ship, as a thing that arrives on a manifest, gets worked, and leaves.

Gwynne Shotwell had been running the operating side of the company for most of two decades by then, and what she had built was the cadence. A launch every few days, boosters flown and reflown until the flying was the cheap part, a manifest that treated a rocket the way an airline treats an aircraft. By the late 2020s the company was putting up the large majority of everything humanity sent toward orbit in a year. The launch was a delivery. What it delivered was the floor.

The economics had settled somewhere in the early part of the decade and held. A booster flew, came back, was inspected, and flew again, often within the week, and after twenty flights the airframe cost had spread so thin that propellant and range time made up a large share of what a launch actually cost. The fairings came down under parachutes, were fished out of the sea, and flew again as well.

What had once been the rare, expensive act of putting something into space had turned into freight. Shotwell turned the manifest into a shipping line schedule, and the schedule was full years out.

The rockets also carried other people's spacecraft: a weather instrument, a group of small commercial satellites or a science probe made by a team with no launch vehicle of its own. A shared manifest let those builders plan around a known ride rather than create a rocket program before they could test an idea in orbit. A returned stage, an inspected engine and another open place on the manifest made the next experiment reachable.

A group buying one of those places could build a receiver with an unusual job. Navigation satellites had been put overhead to tell ships and people where they were. Their radio signals also bounced off the sea. A small craft looking down could catch the scattered return and use its distortion to estimate how rough the water had become under a storm. NASA's CYGNSS satellites had proved that use of the signal; they had also shown how a changing transmitter could spoil the wind estimate until the receiving team learned to correct for it. The next builder would begin with that result, choose a coast or a kind of storm it wanted to understand better, and make its own instrument small enough for an available ride. The rocket's maker would know its mass and the orbit it needed. What the receiver was asking of the sea would be the other team's question.

A slot on a manifest did not answer that question. The group would have to calibrate its receiver, compare a pass with other wind measurements, and revise the instrument if the returned signal wandered. It could spend its working years on that problem instead of on a booster. When the next pass came in, they would place its wind estimate beside a measurement from the sea and see where the two still disagreed.

The satellites themselves came off a line. The company built them by the thousand in a plant that had averaged roughly seventy satellites a week even before the late-decade buildout, each one a mass-produced unit stamped from common parts. A satellite had once been a precious object, designed for a decade, built by hand, flown once, and mourned if it failed. The company made them like a carmaker makes sedans and priced the loss of one into the run. The cadence on the ground and the cadence in the sky were the same fact seen from two ends, a factory feeding a launch site feeding an orbit, all of it moving at the speed of manufacturing. The launch that morning had a control room behind it, a few dozen people at consoles in a building well back from the pad, and most of them had little to do once the count reached zero, because the vehicle flew itself, found its own way back, and landed on its own. The hard human work had moved off the launch and into the factory and the schedule. A flight was the easy part by then.

Other launch programs continued to make their own paths upward. Europe and Japan kept national access to orbit, and China built constellations while working on reusable vehicles. Each program had to solve its own version of the same physical sequence: build the vehicle, clear a range, fly, recover what it could, inspect, and fly again. The number of working ports mattered to the people waiting with payloads. A mission could be designed against a real launch slot rather than an indefinite promise of one.

THE CONSTELLATION AS FLOOR

What sat on that floor, by the late 2020s, was tens of thousands of satellites in low orbit, close enough to the ground that the lag was small and numerous enough that one was always overhead. The optical links were the part that changed the character of the thing. A packet of data leaving a ship in the middle of the Pacific went up to the nearest satellite, crossed the sky by laser from one satellite to the next, all the way over the ocean above the reach of any cable, and came down on the far side near wherever it was going, without ever touching a ground station. The constellation had become a mesh that carried traffic across the parts of the planet that the undersea cables had always skipped.

The constellation renewed itself. A second-generation satellite was built to work for about five years and then to be brought down on command, steered into the atmosphere to burn while a fresh one took its slot. Every week another batch went up and another aging cohort came down, so the mesh overhead never got materially older and was never quite the same hardware twice. A network strung along the ground starts decaying the day it is finished and waits years for a capital program to replace it. The one overhead replaced itself a few dozen satellites at a time, paid for out of a cadence that was already flying.

It reached the ground first through terminals, giving a ship at sea or a settlement beyond the last fiber line a path into the laser-linked mesh overhead. In towns and cities it could add a second route when terrestrial service failed; the dense daily load still ran through fiber and mobile networks. And it reached down to the phone in an ordinary pocket. Direct-to-cell meant a standard handset with no special antenna held a thin connection to a satellite a few hundred miles up, enough for a message from a trailhead or a sinking boat or a town the grid had never come to. The ground carried the heavy traffic; overhead, a second geography joined the coast, the crossing and the high places that cable economics had left apart.

The customers that mattered most were the ones the ground had always underserved. A container ship in the middle of an ocean ran its operations off the constellation, the crew on live video with home, the engines reporting telemetry to a shore office as it happened. Airlines fitted the terminals and sold connectivity at altitude that worked over the poles where nothing else reached. Offshore platforms, mines, pipelines, research stations on the ice, disaster zones where the towers had gone down, all of them set out a box and bought a subscription and had bandwidth within the hour. Places that had been blank on the connectivity map for the whole history of the wired world came online at the speed of bolting down an antenna.

By then the thing carried a fair share of the planet's traffic. Tens of millions of subscribers paid a monthly bill, the bills paid for the launches, the launches put up more satellites, and the whole loop turned on its own without a subsidy. The constellation had taken on jobs no one had built it for. Its signals carried precise time and position, a backup for the aging satellite-navigation systems the whole economy leaned on without thinking. Ships and planes and power grids that had trusted a single fragile timing source for decades now had a second one overhead, denser and harder to knock out.

Filling low orbit with tens of thousands of satellites brought its own problems up with it. The sky near Earth grew crowded, and every object in it moved at miles per second, fast enough that a fleck of debris could hole a spacecraft. The company flew its satellites with automatic collision avoidance, each one nudging itself clear of a tracked object on its own, more than a thousand small maneuvers a day under software control, with human operators supervising the system rather than deciding each maneuver. Astronomers complained that the bright trails crossed their long exposures and smeared the night sky they had studied for centuries, and the company dimmed the satellites and worked the problem with mixed success. A floor laid this thick over the whole planet was a convenience for nearly everyone and a standing worry for the people who had to keep it from colliding with itself. The change an ordinary person noticed was that the dead zones were gone. The drive through the mountains where the phone had always lost its signal, the stretch of interstate across the empty middle of the country, the cabin at the lake, all of them held a bar or two now, routed through a satellite when no tower was in reach. A child born late in the decade would grow up never having watched a parent hold a phone to the sky hunting for a signal.

The reach went where wires had never paid their way. A village in the highlands with no road to it and no copper ever strung toward it put a terminal on a roof and could join the same internet as a capital city, with a connection fast enough to bring its services within reach. Schools that had run on a single shared phone got a classroom of connections. Clinics a day's travel from the nearest hospital sent scans to a specialist and had an answer back before the patient left the room. The wired world had taken a century to build and had stopped at the edge of wherever it stopped being profitable. The thing overhead reached the whole sphere at once, and a good part of the population that had lived outside the network came inside it in the space of a few years.

THE ORBITAL DATA CENTER

A Starlink bus could carry a different kind of payload. Strip away the broadband antennas, give it broad solar arrays, radiators, optical links and a compute module that could be changed as chips improved, and AI1 took shape as an orbital computer. SpaceX designed it for up to 250 kilowatts of peak compute power and 175 kilowatts on average. Its arrays would gather the power, its radiators would shed the heat, and its optical links would connect it to other machines overhead. Bastrop could build the spacecraft again and again.

In Bastrop, Texas, machines were already making Starlink user terminals. SpaceX meant to put AI-satellite production beside that work. A solar manufacturing facility was under construction first, with the satellite production building to follow. The panels grew from solar technology intended for Starlink V3, enlarged to feed a much hungrier machine. The company aimed to send the first AI1s up as soon as late 2027 and build thousands as the line grew. The work would pass from panels and radiators to finished spacecraft, then from the factory to a launch manifest.

The first AI1s would send more than compute into orbit; they would send lessons home. Radiation would test the hardware, sunlight and darkness would cycle heat through the arrays and radiators, and the optical links would have to hold across moving neighbors. Each flight would show what to change in the next production run. A failed unit could be replaced; a better radiator or compute module could take its place on the following launch. Production and launch would let the machine improve while the constellation grew.

The first rung SpaceX set itself was roughly an annualized gigawatt a year of orbital compute by the end of 2027. Beyond it lay a hundred gigawatts a year, then a terawatt, perhaps across as many as a million AI satellites. Each step would need more solar panels and radiators through Bastrop, more completed spacecraft on transporters, more launches carrying them to their places and replacing the ones that wore out. The work at Bastrop could leave as a stream of small machines and gather overhead into something no single launch had carried.

PROPELLANT IN ORBIT

The piece that could turn all of this from a set of destinations into a network was a tank. In the proposed architecture, one variant of Starship would go up, reach low orbit, and stay there as a reservoir of methane and liquid oxygen, a filling station parked in the sky.

Keeping it there was the hard part. Methane and liquid oxygen have to be held at hundreds of degrees below zero, and in orbit the sun beats on the tanks for much of each orbit, driving the propellant toward gas that bleeds away. A depot fought that loss with sunshades and a white skin and a slow roll that spread the heat around the hull, and with active coolers that caught the boil-off and turned it back to liquid before it could get away. Most of the engineering that made a depot work was the engineering of keeping something very cold under a sun that returned each orbit.

Another Starship would launch nearly empty of everything but payload, rise to meet the depot, take on propellant in orbit, and leave for somewhere a fully loaded rocket could never have reached from a standing start on the ground. A ship that fills its tanks after it is already in space can go as far as the depot can keep it fueled, however many top-ups that takes.

The first piece of that architecture had already flown. SpaceX had transferred several metric tons of cryogenic propellant between tanks while in space, a first of its kind operation that gave the company key data for the larger ship-to-ship transfer still ahead.

The tanker flights were unforgiving, and they ran one way. A single ship bound for the Moon or Mars needed its tanks filled to the brim, and filling them took many tanker flights, each one carrying up a load of propellant and coming back empty to go again. A plan that needed ten launches to send one ship was unthinkable when a launch cost a fortune and took months to prepare. It was routine once a launch cost little and flew every few days. The depot was the reason the cadence had to stay high, and the high cadence let the depot pay. Each one made the other make sense.

Kathy Lueders, who ran the Starbase operation that built and launched the ships, was assembling the ground half of a system whose point was to make the ground optional past the first hundred miles. Starbase under her had become a factory and a spaceport in one place, building Starships on a line a few miles from the pads that flew them, turning out ships fast enough that losing one on a test flight cost a week and a lesson, no more. The depots and the tankers and the deep-space ships came off the same line, and the line kept running whether or not any single flight came home.

The depot was the geometric trick that turned a launch vehicle into a transport network. With it, the Moon and Mars and the asteroids stopped being feats and became stops.

The Moon came first, as the proof. The national space agency had bought a version of the ship to land its astronauts on the lunar surface, and that contract turned on the refueling working, because a ship heavy enough to set people down on the Moon and lift them off again could not carry all its propellant from the ground in a single launch. So the first real demonstration of filling a ship in orbit was scheduled around a government flight to the Moon, with the deep-space missions waiting in line behind it. By the late 2020s the depot had a manifest, a date, and a customer who had already paid.

When the first ship-to-ship transfer finally happened, two Starships flying in formation in low orbit, one feeding propellant to the other through a mated coupling while both fell around the Earth at orbital speed, it drew a few minutes of coverage and then slid into the routine. A working depot gave a tug a place to refill while the parts of an orbital yard arrived on other flights. The first loads would be trusses and joints for a platform, followed by solar arrays that powered its tools; later flights could bring tanks and hull sections for a ship too large to leave Earth in one piece. The tug could take each load from its arrival orbit, deliver it to the growing platform, refuel, and return for the next. The work remained in place while rockets came and went. At first every beam would still rise from Earth, and that was the market the asteroid prospectors were scouting when they looked outward for metal that could be worked in space.

All of it bent toward one date on the calendar. Earth and Mars line up for an efficient crossing once about every twenty-six months, a window that opens for a few weeks and then closes for two more years. A depot architecture and a high launch cadence existed in large part to be ready when one of those windows came, to fuel a fleet in orbit and send it all at once while the road was open. The whole apparatus of the orbital decade, the cheap launch and the constellation and the depots and the ships, pointed at the moment a loaded fleet would leave for the next world out. When the window came, the ships would be fueled and waiting.

THE SUPPLY YARD

Before there was a beam to use in orbit, a small probe had to reach a rock. It would ride up with other payloads, leave Earth orbit, match the asteroid's motion, and map the surface closely enough to choose a place to touch. An instrument could measure what the rock held; a sampler could bring back a small piece for a laboratory to test. The first return was knowledge of material and terrain. The next craft could be designed for the rock that had actually been found.

The first commercial wave had arrived at the start of the 2010s. Planetary Resources and Deep Space Industries promised mines among near-Earth asteroids, then were absorbed or wound down before their plans became working extraction systems. Later firms, including AstroForge and Karman+, could use small spacecraft and rideshare launches to test specific pieces of the route. The cheaper seat upward changed the first experiment a small team could afford. It did not itself drill a rock, separate water or deliver metal to a builder in orbit.

Law gave a later cargo one kind of footing. The United States and Luxembourg had drawn rules under which a firm could keep resources it extracted without claiming an asteroid as territory. That right would matter only after machines could collect material, hold it through a trip, and deliver it to someone who could use it. The probe's survey was the first work order for those machines.

At a working orbital yard, a tug would bring a sealed load from a mined body to a receiving berth. A coupling would hold the load while equipment broke and sorted the rough material. Usable metal would move into a forming process; waste and unworked feedstock would stay contained for another pass. Every transfer would take machinery, power and control in a place where a dropped fragment keeps moving. The yard would be a factory as much as a dock.

A finished beam would leave that process on a tug and meet the joints of a growing platform. A crew or robot at the platform would fit it in place beside beams that had come up from Earth. This one had still been found, mined, carried and worked, but it had never ridden a launch vehicle out of

Earth's gravity. The difference would grow with every length of structure made from material already above the planet.

THE PIVOT

Two hundred and fifty-nine days at minimum-energy transit. The next destination was on the ground.

15

Mars and the Kardashev Scale

THE SIMULATOR AT HAWTHORNE

LATE ON A WEEKNIGHT in 2027, in a room at the Hawthorne plant with the lights down, a team ran the Mars-entry simulation for the seventeenth time that day. On the screens a Starship variant carrying a robotic precursor payload fell toward a simulated Martian atmosphere along a trajectory the team had been refining for weeks. The model carried the things that would kill the real ship if they were wrong. Atmospheric density that thinned and thickened with the season and the dust. A terrain map stitched together from years of orbital survey. The way the heat shield burned away on the way down. The thrust profile of the terminal descent, when the engines relit and the ship stood on its own plume and felt for the ground. The logic that picked a landing spot in the last few seconds out of whatever the cameras saw. The room was quiet. The simulation ran in compressed time, seconds standing in for the long minutes of the real descent, and the simulated Starship came down on the simulated plain and held there on its legs. Someone rolled the seed. The simulation ran again.

From the back of the room it looked like office work. A dozen people at workstations, the glow of the screens on their faces, the low talk of engineers reading numbers off to one another, a plot of the descent redrawing itself each time a run finished. There was coffee going cold and the particular quiet of people concentrating late. What they watched was a curve, the line of a trajectory bending down through a column of simulated air, with a scatter of numbers beside it that told them whether the ship in the model had lived or died. The work of getting to Mars, at this stage, was the work of making that curve come out right a thousand times over before anyone bet a real ship on it.

Descent remained one of the hardest parts of sending a large craft to Mars. The planet's air was thick enough to tear at a spacecraft and burn it on the way in, and thin enough that it gave a heavy ship almost nothing to brake against near the ground. Parachutes had helped land smaller probes; a ship this large would need much more braking than a parachute alone could give. Every probe that had landed had threaded a narrow window between coming in too fast to stop and too slow to hold its heading, and the larger the ship the narrower the window got. A Starship would attempt the descent at a scale far beyond the robotic landers that had gone before. The simulation was the team's way of flying the landing a thousand times in software before they flew it once for real, failing in the model where failing was cheap, until the trajectory that brought the ship down was one they had already watched survive ten thousand times.

The heat was its own problem. A ship hitting the Martian air at interplanetary speed pushed the gas in front of it into a sheath of plasma hot enough to glow, and the side facing the wind had to be armored to take it without burning through to the structure beneath. The shield that did the work on the way in was dead weight the rest of the time, mass carried all the way from Earth and across the gap to be used hard for a few minutes and then ignored. The simulation tracked the heat as carefully as the trajectory, because a shield that came apart a minute early ended the mission as surely as a leg that folded on landing.

The failures in the model were specific and various. The ship came down on a slope and tipped past the angle its legs could hold. It planted a leg on a boulder the cameras had read as flat ground and rocked over. It burned through the last of its descent fuel a hundred meters up because a gust of thin wind had pushed it off the line and it had spent the margin correcting. It picked a clear patch in the final seconds and found a field of rocks just past the edge of the frame. Each way of dying taught the team something to fold back into the next run, a wider leg stance, a sharper eye for hazards, a little more fuel held back for the last bad surprise.

And the landing had to fly itself. By the time a ship reached Mars the planet could sit a dozen light-minutes away or more, far enough that the whole descent, the fire and the turn and the final settling onto the legs, would be over and the ship alive or dead before the first picture of it entering the air had even reached home. A controller at Hawthorne could watch and could not steer. The landing had to make its own decisions, reading the ground through its own cameras and firing its own engines on its own judgment, in the last seconds, with the nearest human help a quarter hour behind the news. The seventeen runs were teaching a machine to fly the landing alone.

The early manifests carried machines before people, the cargo of a settlement that had no settlers yet, landed and waiting. Power units. Propellant plants. Water-mining gear. The robots to set it all up. A crew that flew later would arrive to find fuel already made and power already running and a stock of supplies already on the ground, because sending people to a bare planet and asking them to build all of that by hand was how expeditions died. The plan was to land the hard parts first, prove they ran through a Martian year with no one there to fix them, and only then put a human being on a ship.

The machines that went first had to do the work a construction crew would normally do, with no crew there. A rover would scout the ground the orbital maps had picked and confirm the ice was where the data said. Diggers would break it out and haul it to the plant. The plant would stand itself up, unfold its panels, and begin the slow business of turning air and ice into fuel and oxygen and water. All of it had to work through a year of cold and dust on commands sent from a control room on Earth, every instruction arriving minutes late, every problem solved by a machine that could not phone home and wait for an answer. Everything humanity put on Mars in those first years, it put there to run alone.

Bill Gerstenmaier had spent much of his career at NASA, including years responsible for human spaceflight, before joining SpaceX. A vehicle meant to carry people would need more than a descent that survived in software; its models would have to be checked against hardware tests, instrument readings and flights before a crew relied on them. The seventeenth run of the night could identify a failure worth carrying into the next test.

A simulation could be reset after every failure. The team could change one parameter, run the descent again, and compare the new trace with the old one before a real ship took the risk. Hardware would answer questions the model missed; the next model would have to absorb those answers. The repeated runs made a place for failure early in the work, where it could still alter a shield, a fuel margin or the landing logic. The room stayed at its screens while the curve redrew itself.

THE METHANE ON MARS

What made the descent worth simulating was the climb back out. A ship that lands on Mars and cannot leave is a probe. A ship that lands and flies home is the start of something else, and the whole distance between the two is propellant. No ready-made return propellant waited on Mars, and no ship could carry enough from Earth to land with fuel still in the tanks for the return, so the ship would have to make its own on the ground.

The chemistry was old. A French chemist named Paul Sabatier had worked it out before the First World War, running carbon dioxide and hydrogen together over a catalyst to get methane and water. Mars handed the engineers the carbon dioxide for nothing, since its thin air was almost all of it. The hydrogen came from water, and there was water on Mars, frozen into the ground across much of the planet. Pull the carbon dioxide out of the air, split the ice into hydrogen and oxygen, run the hydrogen and the carbon dioxide together over the catalyst for methane, and set the oxygen aside. Methane and oxygen were the two things a Starship burned. The return propellant would be made on Mars, by a chemical plant set down on the surface and switched on, out of the air and the ice that were already there.

Making it took power, and a great deal of it. To fill a single Starship for the trip home, a plant had to pull tons of carbon dioxide out of a thin atmosphere. It had to melt and split tons of buried ice for the hydrogen. It had to run the chemistry and chill the results down to liquid. All of that ran on electricity, and Mars did not hand out electricity for free. The plan called for fields of solar panels unrolled across the Martian ground and, in some versions, compact nuclear reactors set down beside the plant, enough generating capacity to run a small industrial town. The first thing a settlement would build was its power. Everything else, the fuel and the water and the air to breathe, came out of the power.

The dust was the reason the reactors kept turning up in the plans. Every so often a storm rose on Mars that did not stay local, that spread until it wrapped the whole planet and hung there for weeks, throwing enough fine dust into the sky to dim the sun across an entire hemisphere. A settlement that ran on sunlight alone would watch its power fall away just when it could least afford to lose it, its panels graying over under a fall of dust no one was there to sweep. A reactor did not care whether the sun was out. For a base that had to keep its people breathing and its fuel cold through whatever the planet did, that indifference was worth a great deal.

The water was the part that tied the plant to a place. Carbon dioxide filled the Martian air everywhere on the planet, while water ice lay buried only in certain ground, and a propellant plant had to sit where the ice was, near enough to mine it and haul it without spending more energy on the moving than the ice was worth. So the choice of where the first ships came down was made years ahead, off orbital maps that read the ice through the soil, and it was made by chemists and miners as much as by pilots. A precursor would go first and prove the whole chain on a small scale, pull a little air, melt a little ice, make a little methane, hold it through the cold and the dust, and report back that the cycle held. Until a robot had made fuel on Mars and kept it, no crew was going anywhere.

The ground chosen for water would invite a second kind of work. A drill sent down to find ice could bring a core up before the material went to the plant. Its bands of ice and dust would have to be kept in order, measured and compared with the layers the orbiters had seen from above. Martian ice can hold a record of changes in climate; orbiting radar has already found such layers in the polar deposits and mapped buried ice at lower latitudes. A crew on the ground could ask whether the deposit under its own site formed in the same way, how often dust interrupted its growth, and what those changes say about the planet before anyone lived there. The first answer might be that the orbital map had read the site badly. Even that would be a discovery the next landing could use.

The plant would still need most of the ice for fuel, water and air. The core would not stop the diggers. It would change what the diggers noticed. A layer that made extraction harder might mark a chapter of Martian weather no spacecraft had been close enough to read. The people who crossed the distance to keep a settlement alive would be the first people able to return to that cut in the ground, take another sample and learn whether the first one was typical. Their familiarity with a place would begin with the work that fed them, then grow beyond it.

The scale of it was hard to hold in the head. For the return, one ship could need about a thousand tons of locally made propellant before it could leave Mars. Making methane and oxygen on Mars at that scale would mean running a plant long before the ship was ready to leave, through the cold and the dark and the long dust storms. A plant that size was a small power station and a chemical works and a mine, all of it landed in pieces and assembled by machines, all of it expected to run for years with the nearest repair crew on another planet.

The plant made more than fuel. It made oxygen too, and a settlement needed oxygen for more than burning. It needed it to breathe. The machine set down to fill a rocket was the same kind of machine that would keep a habitat alive, and the line between an industrial plant and a life-support system, so sharp on Earth, ran together on a planet where the air outside could not be breathed.

The choice of methane went back to the same planet. The ships could have flown on the fuels rockets had always used, the hydrogen that burned cleanest or the refined kerosene that packed tightest, and on Earth either would have served well enough. Mars decided it. Hydrogen had to be kept colder than almost anything and slipped through every seal given the chance, and no one was going to refine kerosene on a world with no oil. Methane could be made on Mars out of the air and the ice, held at temperatures close enough to the oxygen that the same kind of cryogenic tanks and plumbing suited both, and burned in an engine clean enough to fire again and again. The engine that powered the whole system had been designed around a fuel chosen for a planet most of its flights would never visit. The destination had reached back and shaped the machine.

By the late 2020s the process sat in the mission plan as a line item, a capability to be procured and proven out like any other. The whole round trip turned on it. A Starship that could fuel itself for the journey home was a Starship that could come home, and once one ship could come home the trip stopped being an expedition and started being a route.

The buildout fell across more than one of those round trips. The first window would send cargo only, the precursors and the first plants, to a chosen patch of ground. The next would send more machines and, if the plants had held, the first people, who would come down to a base the machines had already started. Each window after that could land more, because each one left more behind. The place would accrete, one launch window at a time, the way a port grows up around the first dock, never arriving whole on any single ship.

Fuel and air and water were only the first things a settlement would have to make for itself. A base that shipped everything else from Earth would stay an outpost, dependent on a supply line that ran across the better part of a year and only opened every other one. The long aim was to make more and more on Mars over time, the food and the parts and eventually the machines themselves, until the place could lose its link to Earth for a while and live. That was the line between a research station and a settlement, and crossing it was the work of decades, measured in lifetimes more than in launch windows.

THE CYCLER

In the latter part of the twentieth century the astronaut Buzz Aldrin, the second man to walk on the Moon, had worked out a trajectory he called a cycler. A vehicle placed on the cycler orbit would loop forever between Earth and Mars along a fixed path, swinging close to one world and then falling away toward the other, never stopping, never landing, carried around the loop by gravity and momentum alone. Once it was moving it cost almost nothing to keep it moving. The fuel went into the small ferries that ran up from each planet to catch the cycler as it passed and dropped back down when they arrived.

What made the idea worth reviving was what the journey did to people. Six months or more in deep space exposed a crew to radiation the entire crossing, cosmic rays and solar storms that a thin-walled ship did little to stop, and the dose added up over a career to something no space agency would let its astronauts take lightly. Shielding was heavy, and every ton of it was a ton that had to be launched and pushed across the gap and slowed down at the far end. A cycler carried its shielding once and kept it, a heavy and well-protected vehicle that paid the launch cost of its mass a single time and then used that mass again on every pass for decades. The ferry that ran up to meet it could stay light, because it only had to survive the short hop to orbit and the short hop down.

Time was the other cost. The crossing ran six to nine months each way depending on the year, and a body in weightlessness for that long lost bone and muscle and more, no matter how many hours its owner spent strapped to a treadmill. A vehicle that crossed often enough could be built to spin, giving its crew some pull underfoot and landing them in better shape than a still ship would. One that made the trip again and again could carry comforts a single-use ship could never spend the mass on. The case for the cycler was partly a case about what the journey took out of the people who made it.

And once a crew left, it was committed. A mission to the Moon could turn around and come home in a few days if something broke. A mission to Mars could not. Past a certain point a few weeks into the burn, the cheapest way back to Earth ran forward, around Mars and onward, a loop of many months with nowhere to stop. Everything that kept the crew alive had to keep working all the way, with no harbor to limp back to, on a ship that was its own hospital and its own repair shop and its own lifeboat. The distance changed what reliability meant. On a Mars mission it was the difference between a crew that came home and one that did not.

Bret Drake and other Mars-architecture planners had studied the trajectories, launch windows and gravity assists that a recurring route would require. What the math described was a piece of standing infrastructure. A cycler turned the trip to Mars into something scheduled. The ferry left when the cycler came around, the way a train leaves when it reaches the platform, and the great mass of the transit vehicle, its shielding and its living room and its stores, stayed up in the loop and was used again on the next pass and the pass after that. The expedition became a timetable.

The loop came almost free, and Drake's math was largely about the almost. A cycler left to itself would drift off true as the planets tugged at it unevenly, so it had to be set up to swing close past Earth or Mars at the right moments and let their gravity bend it back onto the path, with only small burns to trim what gravity left undone. Getting that to close, so that the same orbit came back around to meet both planets year after year, was the hard and beautiful problem under the whole idea. It was orbital mechanics worked like a piece of clockwork, every gear a world.

The cycler would come last. The early missions would fly the hard and direct way, each ship carrying everything it needed and burning most of it to get there, because a cycler only paid once there was enough traffic to fill the ferries that fed it. The cycler was the shape of the mature route, the thing the traffic would justify after the first decade had shown the planet could be reached and lived on at all. It sat in the plans the way a rail line sits in the plans of a mining town, drawn before the ore is found, waiting for the day the wagons can no longer keep up.

The solar system set the timetable. Earth and Mars came into the right alignment for the crossing only about once every twenty-six months, when the two planets sat on the same side of the sun close enough that a ship could make the jump on a reasonable amount of fuel. Miss the window and the next one was more than two years away. So the whole enterprise ran on that cycle. Ships and crews and cargo were readied against a date the solar system handed down, and fleets went out in waves when the window opened and then the sky closed for two years while the planets drew apart and came back around. A program built on that cadence answered to the planets before it answered to money or will. When the window came it was ready or it was not, and there was no buying the difference.

THE KARDASHEV FRAME

There was an old scale for thinking about the energy a civilization could command. In 1964 a Soviet astronomer named Nikolai Kardashev, looking for a way to rank the civilizations he hoped to detect across interstellar distances, proposed to sort them by the sheer quantity of energy they could command. His first kind could command power on the order of the Earth of his day. The second drew on the output of a star. The third commanded a galaxy's energy. The gulf between the rungs was immense. For decades the scale lived mostly in speculative astronomy, a way of thinking about signals from very far away.

Kardashev had reached for energy because energy was the one thing that showed at a distance. A civilization's art and its politics and its sciences stayed local, invisible across the light-years, while the heat and light it threw off could in principle be seen from another star. Energy was the measure that survived the distance. It also happened to be the measure that mattered on the ground, because everything a civilization could do, it did by moving energy around, and a species that could command more of it could do more.

Humanity sat well short of even the first rung. In the later, widely used version of the scale, the first kind stands for a planetary energy budget far above present human use. The human world still drew much of its power from the buried remains of old sunlight while beginning to take more of the living sunlight directly. Carl Sagan, who did as much as anyone to carry the idea to a wide audience, later smoothed the three steps into a continuous scale, so that a world could be rated by a fraction of the way to the next tier, and on that scale humanity in the middle of the decade would rate a little over 0.7, partway up the first slope, still mostly burning what it dug up.

The lower reach of the first rung was visible in ordinary equipment. A solar field gathered an afternoon's light; a battery kept part of it for evening; a propellant plant on Mars would need to turn local ice and air into the fuel for a return flight. SpaceX had named Type II, the full output of the Sun, as a distant aspiration. The nearer work was learning to gather, hold and use more energy where a task required it, then making those operations repeatable beyond one site or one world.

At the far edge of the same frame, Musk imagined making solar arrays and radiators for orbital compute on the Moon, then launching them from its airless, low-gravity surface by electromagnetic mass driver. The ladder would reach past gathering sunlight to making the machines that caught it on another world.

The scale was a crude instrument, and no one who used it pretended otherwise. A civilization was more than the energy it burned, and a number on Kardashev's ladder said nothing about whether the people on the rung were wise or free or content. What the scale was good for was the long view, the question of what a species could reach for at all, and on that one axis the answer had started to change. A world that could make energy on more than one planet had more it could attempt than a world that could not, whatever it chose to do with the difference.

THE HEDGE AND THE THESIS

The case for going had two forms, and both could be true at once. One was a hedge against ending: a species living on one planet can be reached by one planet's catastrophe. The other was a case for growth: another world could offer new ground for making energy, materials and knowledge. Musk had argued both in public for years. Neither motive filled a tank or made a landing safe, but both gave a reason to do the work that could.

The hedge had a fossil record behind it. The rock that ended the age of the dinosaurs had been one rock on one ordinary day, and it was not the only such ending the record in the ground remembered, whole chapters of life closed out fast. A species kept entirely on one world was a species betting that no such morning would come while it was the tenant, and the bet had to be won every year forever. Spread the species over two worlds and the same morning could not finish it. People had been making that case, in one form or another, since long before anyone could act on it.

The growth case was harder to feel and larger in the end. A species kept to one planet had a ceiling it could see, the fixed budget of a single world's sunlight and rock, and sooner or later everything it wanted to do would press against that edge. A species that could reach past its own planet had no such fixed edge in sight. The first case was about not dying. The second was about how much there was to do if you lived.

A refuge on Earth would be easier to build and reach than one on Mars, even after a severe disaster. Money and skill used for a distant settlement could also protect lives on the planet people already inhabit. Mars would make the price of any second home explicit: power for every breath, a plant for every return trip, and parts that could be repaired without a quick supply run. Those demands belonged inside the ambition as design work still to be done.

The same ships, power plants and propellant works would serve both motives. People would have to build and maintain them.

The people who would go first would be builders before they were anything else. The robots would have gone ahead of them, so the crews would land into the start of a settlement and spend their years adding to it, putting together the first place humans had ever lived off the Earth. They would stay through a Martian year or more, because the window home opened only on the solar system's schedule, and the early ones might wait a long time for a return that was never guaranteed. They were signing up to be the first humans to live somewhere the species had only ever seen through a telescope. Whatever the arguments said, somebody was going to have to go and find out what it was actually like.

16

The Stack Outlives the Founder

WHAT NOW EXISTS

AT CAPE CANAVERAL, A Falcon 9 launch crew can send a vehicle into the morning with engines that have already been fired on stands in McGregor. The flight sends back measurements while the first stage separates, turns and makes for a landing place. When that stage returns, a crew secures it, takes it to a hangar, and checks what the heat and vibration have done before it flies again. A launch is a particular morning of weather, hardware and people. The process that surrounds it survives the morning. The next vehicle is already taking shape while the engineers read what this one brought back.

That process has acquired an extraordinary reach. Falcon boosters have come home and flown again often enough that returning one is part of an ordinary work order. Starlink satellites come off a production line, ride up in batches, find their places in low orbit and hand a signal across one another to a terminal on the ground. A Megapack leaves a California line with cells, cooling and controls already inside the cabinet, ready for a crew to connect it to a grid. A casting press closes its die, receives a shot of molten aluminium, and opens on a part that replaces a long welded assembly. Each of these is a working operation, not a drawing of one. The number of repetitions has changed what a launch, a connection, an evening's stored sunlight or a car body can cost.

Other layers are at different stages. The robots in Tesla's factories are still learning jobs under the eyes of technicians. The proposed chip-manufacturing venture in Grimes County has a site and a framework, with its commercial terms and first production still ahead of it. Orbital compute remains a design that has to prove its power, radiators and communications in flight. Their place in the stack is the next set of interfaces being attempted, where a line already learned to make one hard object may become a way to make another.

The connections that operate now are already consequential. A cell developed for a vehicle can be packed into a cabinet that steadies a grid. Falcon launches carry customer cargo and loft Starlink satellites; those satellites can relay telemetry from a Starship test flight. A camera fleet on roads gathers the awkward turns and near misses that become the next training cases for a driving system. The handoff is physical at one end and informational at the other, a part with a test record, a stage with a flight log, a clip of an intersection taken back to an engineer. Across those handoffs, a gain made in one place can be used somewhere else without beginning again from raw metal or an empty screen.

The earliest pieces were built inside one working career. SpaceX was founded in 2002; the first Falcon 1 reached orbit only after three failed attempts. The first large Tesla car factories came later, and the grid-storage line and training chips later still. Some of the people who worked on the first layers were present to see the next ones arrive. They had watched a booster landing turn from a wreck on a ship's deck into a morning when a recovery crew tied down a scorched but intact stage. They had watched a small battery cell become a pack, then a cabinet, then a row of cabinets whose response to a grid fault could be measured in milliseconds. What accumulated was a set of ways to repeat difficult work, with the tools and the people to keep repeating it.

WHAT CAN BE HANDED ON

Elon Musk set a great deal of this in motion and held unlike things near enough to one another that their builders could borrow. An engine team, a satellite line, a car plant and an AI group did not naturally share a timetable. Decisions made at the top sometimes forced a connection before the organizations beneath were ready for it. The speed came partly from one person willing to make calls across company boundaries, and partly from teams able to carry them out. The question at the end of this book is what remains available when the decision has to be made by people who were not there at the start.

On a working line, an answer is already being written. The die for a cast underbody holds a decision about alloy and cooling in hardened steel. Beside it are shot records, scans of rejected parts, measurements of wear and the maintenance plan for the next shift. A person who did not design the first die can read the evidence of the failures that shaped it, watch a fresh casting come out, and decide whether this run is still inside tolerance. In McGregor an engine test does much the same for a Raptor that will later hang under a booster. The builders leave traces another crew can use. A process becomes portable when the next crew can make a sound part, find a bad one, and improve the line without waiting for its first author to return.

The people have already moved. JB Straubel carried the cell and pack problem through Tesla's early years and later turned to recovering the metals from cells whose first lives were finished. Engineers who worked on the cars' vision and the first in-house training chips left for other projects. Those departures tested different parts of the organization. Gwynne Shotwell had spent the better part of two decades running the launch business, through crew flights, customer launches and the long campaign to bring boosters home. Her continuing leadership was another route to continuity: an operator who stayed, kept the work moving and helped others learn it. A line that kept shipping after its first builder left had acquired trained successors, equipment, records and suppliers that could carry the work; a new company founded by someone who had learned the work could carry part of it outward.

The copied thing is rarely a whole company. It may be a test protocol, an alloy family, a way of laying out a factory bay so a worker can reach the part, or a habit of sending a failed unit back to the people who can change the next one. Those travel through people, drawings, supplier orders and the ordinary pass-down from one shift to another. Around El Segundo, where the first SpaceX engineers worked in rented industrial space, later hardware firms have placed design and fabrication close together for work of their own. They inherit neither the early rocket nor the exact people who built it. They can inherit the practice of letting the maker and the designer see the same failure while there is still time to change the next unit.

The appetite that joined so many layers under one founder cannot be issued as a procedure. Someone still has to choose which hard middle step is worth years of work, when to keep a line running and when to tear it out, and how much to spend before an instrument gives a clear answer. That judgment belongs to people, with their risks and their limited working lives. The plant, the test stand and the flight log make the decision less dependent on a single person's memory. They allow another builder to begin from the last sound result and carry the work into a place its first builder never planned to go.

THE WORK STILL AHEAD

A sequencer can read a sample and return a string of information about living material. A cell-culture line can grow material whose behavior must be watched and measured again after every change of conditions. Between a finding on one bench and a treatment made reliably for many people lie the vessels, instruments, records and trials that show the same result can be produced without hiding a dangerous variation inside the batch. The work has its own clock. Cells grow at the rate cells grow; regulators ask for evidence across time and across patients because a body has to live with the result. The possibility is large precisely because the next step is so exact. If a process can hold a useful biological result steady as the runs grow, a discovery made for one case can become care that reaches beyond the laboratory that found it.

The methods are already present in pieces: gene sequencers, cell lines, instruments that read what a culture has made, factories that can grow material under controlled conditions, clinical teams that test the outcome. Their handoffs are harder than an engineer's drawing of one seamless line makes them look. A measurement made on a small culture has to remain meaningful when a larger vessel changes the heat, mixing and timing of the growth. A production record has to follow the material into a trial and, if the evidence holds, into ordinary use. This is work for biologists, process engineers, clinicians and the people who maintain the instruments. A process engineer watching the larger vessel and a biologist who knows the small culture need to see the same change, though a living system will answer back in ways a metal casting does not.

A material begins even smaller. Someone makes a first gram and gives it a name because its properties are useful. In the next run the temperature drifts, the mixture leaves a trace of something unwanted, or the cooling changes its structure. The result that made the first sample exciting disappears as the process grows. A team that can measure every batch, trace the change to a step in synthesis and alter that step has begun to turn a discovery into something another workshop can buy and use. The hundredth run matters as much as the first. A new material that keeps its strength, conductivity or durability when made by the ton can leave the bench and become a part in machines whose designers were previously confined to the substances already in the catalog.

The long middle of that work is easy to overlook from the outside. There are fixtures to build, instruments to calibrate, reject material to examine and a yield that rises slowly as the people on the line learn which readings warn them early. The same sequence runs through a die, a cell plant and a chemical works, though the substances and tolerances change. A useful material at volume would give later builders a new starting point, the way cheap lithium-ion cells gave the car and the grid a part that neither had first invented for itself. The next use might be found by someone who never met the chemist who made the first gram.

In a classroom the test is different. A child is learning to scale a drawing. She changes one length and leaves another as it was, three attempts in a row, enough to show the teacher where understanding broke before the next lesson runs past it. A digital tutor could keep the attempts in order and offer a different explanation or a simpler next problem. The teacher could see the same record and ask the child to show how the answer was reached. A wrong answer at the end of a page says little; the same step repeated gives them something to work on together. The models now being tested for teaching do not settle curriculum, trust or access. A school would still have to decide what to teach and whether the instrument was helping.

The next day the child could try the problem again with the mistaken step visible beside the new one. If the method now holds, the teacher can offer a problem that does not fit the old pattern and ask the child to choose an approach. After class the child opens a drawing of a small bridge she wants to build across a tabletop river. Its span has to match the scale of the banks she made the week before. She sets a length on the drawing, works out where the supports would stand and cuts the first piece for the model. It may still fall when she puts it together. The next calculation is hers to make.

The planet produces signals on another scale. A sensor reads water or heat at one place; a satellite sees a wider pattern; a forecast draws on both and offers a farmer, a utility crew or a town time to act before a condition arrives. Each link can already be built in some form. Making the information dependable, timely and usable where decisions are made is the harder systems problem. A town with a better picture of the water approaching its streets has to connect that picture to people who can open a gate, clear a drain or move equipment. An energy system has to connect a forecast of sun and wind to the storage that can carry power into the hours after the weather changes. The sensors do not run the planet. They let people meet a larger piece of it with a little more knowledge and time.

Carbon drawn from the air by a pilot plant is another small unit whose scale is easy to pronounce and difficult to build. Its fans, capture material, heat supply and measured output must survive repeated cycles, and the energy to run the process has to be counted with the carbon it removes. That accounting belongs to the apparatus itself. A plant that can be repeated without spending more energy and material than the work is worth would offer one tool among many for tending a shared atmosphere. The land, oceans and air will never be one operator's factory floor. They are the ground on which every factory, field and home depends, and building a useful instrument for them means leaving a record that other places can test against their own conditions.

Each job is measured by what happens at the far end of the apparatus. A clinical approval takes the time needed to see what a treatment does; a material must hold its properties through many runs; a learner has to actually learn; a forecast has to arrive early enough and accurately enough to change a decision. The people who build the instrument and the people who use it have to stay in contact through those cycles. The long reach comes from repetition that survives correction. One sound batch, one understood lesson, one avoided grid failure or one better-timed response to weather becomes the work that can be done again, in another building and by another crew.

THE METHOD IN OTHER HANDS

CATL's cells reached vehicle and storage makers outside its own walls; BYD showed another route by bringing its own cell work into the car body. Wang Chuanfu and Robin Zeng ran unlike organizations, and each left later builders a different starting point. A qualified cell could enter a machine its maker had never drawn.

The frontier AI laboratories were finding the same dependency from the opposite end. A model lived in software, yet training the next one meant placing chips in racks, carrying enough current to them, moving their heat and feeding them data in a form they could use. Anthropic, OpenAI and Google DeepMind had different models and different arrangements for obtaining those physical layers. Some designed chips; some rented large blocks of compute; all needed power and a place where the machines could run. Their choices pushed new work toward semiconductor plants, data-center crews, utilities and builders of cooling systems. The advance in a model was real, and so was the column of physical work underneath it.

Money enters long before any of those lines return a useful part. A giant casting die must be cut before the first saleable underbody is poured. A transformer must be ordered before a cluster can take its full load. A launch tower stands through testing and delay before it catches a booster. These are years of materials, skilled work and equipment held against a future rate of use. The financing matters because it lets a builder keep the crew and the apparatus together through those years. Its measure in the story is the process that finally repeats, a casting that passes inspection, a rack that can be energized, a booster caught and prepared for another flight.

THE WORK THAT TRAVELED

The launch schedule built to sell rides to customers lifted the company's own satellites, and the rides it sold gave other operators a way into orbit on a timetable they could plan a business around. The cells developed for cars moved into storage, first for homes and then for grids, until battery production served a substation as readily as a driveway. The vehicle factories gave robots a working floor on which to learn jobs beyond the vehicle line. Once the machine worked, its builders could find uses for it beyond those in the first drawing.

The work had left its marks in a returning stage, a terminal on a roof, a cell in a cabinet, a casting in a car, a hand learning to close on an object, and a vehicle choosing a path through a street. The plans beyond them, an orbital computer or a plant making propellant on Mars, would need their own years of proof. The existing lines could supply parts and habits for those attempts, while new builders would have to find the failures that appear only when a machine is asked to do that new work. At the end of a shift there is usually no ceremony for any of this. A record is signed, a part moves on, and someone at the next station takes it up.

THE IMAGE AT THE CLOSE

Follow a Megapack out of the Lathrop factory. It is a gray steel box large enough to need a flatbed, with cells, power electronics, cooling and controls packed for installation at a grid site. At the gate the driver checks the delivery address on a screen in the cab and pulls the trailer into ordinary traffic, past warehouses and fields. At the destination a crew will set the cabinet on a prepared pad, make the electrical and data connections, test the circuits, and hand it over to the grid operator. On a bright afternoon the operator's schedule calls for a charge. A command reaches the cabinet over a data link; switches close, current enters the cells, and coolant carries off the heat. When evening demand rises, the schedule calls that power back down the cables.

The flatbed has gone on to another load. Back in Lathrop, another cabinet is moving through the line.

A Note on Sources

This book inspects the companies from outside the fence. It does not rely on interviews granted for the book, on access to any company facility beyond what the companies show the public, or on documents unavailable to a reader willing to look for them. That boundary was a choice as much as a constraint. A public case can be checked by any reader who doubts it. More important, a book arguing that the pattern lay in plain sight had to build its case from what lay in plain sight.

The 2026 filing record forms the spine of the book's company account. SpaceX's registration statement, its June 5 prospectus and subsequent public updates give dated evidence for claims about the consolidated business. The compute agreements with Anthropic and Google, the audience and usage figures for X and Grok, the February 2026 absorption of xAI, the framework around Terafab and the orbital-compute risk factors come from that record. An announced project remains a plan on these pages even when its engineering is described closely.

Other public records surround that filing. Launch logs, webcast archives, and published actions by the Federal Aviation Administration date the flights, the losses, and the reviews. Earnings calls, investor presentations, and technical talks in the companies' own voices supply the engineering detail for the factory, battery, and autonomy chapters. Those chapters also draw on Munro and Associates' published teardown work and on the filings and disclosures of competitors and customers discussed in these pages. The rest comes from the daily record of the trade and financial press, which covered each company on its own beat, thoroughly and well. This book gathers what that coverage established, then stands back from it. It could not have been written without that work.

The book moves between documented events and illustrative scenes. Accounts of launches, tests and public demonstrations draw on footage, photographs, company tours and published accounts. The supervised turn and the four-site October day set known machines to work in constructed circumstances; their clock times and people belong to those scenes. The Tennessee deer encounter is an illustrative construction of the fleet-data mechanism, not a documented drive with a known clock time, location, driver or uploaded clip.

The numbers are treated as dated measurements. Roughly, about and on the order of mark estimates with real uncertainty. Costs, prices, counts and capacities move quickly in these industries. Most operating figures in the book come from records available by mid-2026; the AI1 design and schedule in the orbital chapter use SpaceX's later Starmind description available on 6 October 2026. A reader checking them later should expect values and plans to have moved. The notes carry the available source trail for individual claims.

Part IV carries the constructed-scene method farther forward. The Austin morning, orbital yard and Mars simulator show possibilities opened by working machines, announced plans and physical constraints. They do not promise those particular outcomes. The closing chapter returns to the work already in hand and the people who may carry it forward.

The notes that follow carry the accounting, chapter by chapter, keyed to the text by phrase. The deeper debt is to the reporters who built the record this inspection walks through, beat by beat, company by company, year by year. The book's argument is that almost no one stood back far enough to see the machine whole. The record that made standing back possible is theirs.

Notes

CHAPTER 1
FRIDAY AT STARBASE

3 "Starship Version 3 stood on Pad 2": SpaceX, Starship Flight 12 mission record identifies the first V3 flight from Pad 2.

3 "At 5:30 p.m. Central Time, it would lift off": SpaceX, Starship Flight 12 mission record gives liftoff on 22 May 2026 at 5:30 p.m. Central Time.

4 "the first full-stack liftoff tore into the launch mount": Federal Aviation Administration, *Written Re-evaluation of the 2022 PEA for Starship/Super Heavy* (November 2023), records disintegration of the concrete pad during the 20 April 2023 flight, debris deposition and the redesigned foundation, steel plating and water cooling: https://www.faa.gov/media/72826.

4 "Crews filled the site that day": SpaceX, *Starbase Overview* (2024), reports more than 1,800 direct Starbase employees: https://www.spacex.com/vehicles/starship/assets/media/Starbase%20Overview.pdf. That site-level count does not establish a launch-day headcount.

6 "Starbase had become an incorporated Texas city": Cameron County, *City of Starbase Incorporation Election*, 3 May 2025 county results (the linked summary is labeled unofficial) report the incorporation vote: https://www.cameroncountytx.gov/elections/wp-content/uploads/2025/05/City-of-Starbase-Summary.pdf. The present city government confirms incorporation: https://www.starbase.texas.gov/city-commission-and-staff.

6 "All thirty-three Raptor 3 engines on the Super Heavy booster ignited": SpaceX, Starship Flight 12 mission record records ignition of all 33 Raptor 3 booster engines.

8 "one of the booster's thirty-three Raptors shut down": SpaceX, Starship Flight 12 mission record records one booster engine shutting down during ascent and one upper-stage vacuum engine loss.

9 "At McGregor, in central Texas": SpaceX, June 2026 prospectus, pp. 282–283, describes Raptor manufacture at Hawthorne and qualification, acceptance and post-flight Raptor testing at McGregor, including dedicated vertical stands: https://www.sec.gov/Archives/edgar/data/1181412/000162828026041013/japanfwp_06042026.htm. The passage does not claim that every flight engine is test-fired there.

9 "driven toward manufacturing as hard as performance": Elon Musk, walking tour of SpaceX Starbase with Tim Dodd (Everyday Astronaut), Boca Chica, Texas, 2021; Musk frames Raptor's hard problem as cost per ton of thrust below $1,000 and puts the work of the manufacturing system orders of magnitude above the work of the design.

10 "with Intel's technical help": SpaceX Form S-1, May 2026, as amended; the filing describes the Terafab initiative together with Tesla, with Intel joining the project in early April 2026. The passage describes the proposed Terafab project, not an operating chip factory; the work listed is what making circuits there would require.

11 "the Federal Aviation Administration designated the flight a mishap": Federal Aviation Administration, general statements (27 May 2026) identifies the Flight 12 mishap and investigation requirement.

12 "released twenty Starlink mass simulators and two modified Starlink satellites": SpaceX, Starship Flight 12 mission record records 20 mass simulators and two modified Starlink imaging satellites.

13 "The first flight was lost to a fire": NASA, *Compendium of Launch Vehicle Failures* (2017), pp. 48–50, reviews Falcon 1's first three failures, including first-stage fire, second-flight propellant slosh following stage contact, and third-flight residual-thrust recontact: https://ntrs.nasa.gov/api/citations/20170009844/downloads/20170009844.pdf. Musk described the fourth attempt as the program's last funded Falcon 1 launch in his 2017 IAC talk: https://www.spacex.com/media/making_life_multiplanetary_transcript_2017.pdf.

13 "lost on the pad during a routine fueling test": SpaceX, "AMOS-6 Static Fire," 1 September 2016, records the vehicle and payload loss during a standard pre-launch static fire test: https://www.spacex.com/updates/reusability.

14 "tore itself apart over the Gulf": SpaceX Starship flight-test records; the first integrated Starship test flight, April 2023.

14 "caught the descending booster clean out of the air": SpaceX Starship Flight 5, October 13, 2024; the launch tower's mechanical arms made the first successful catch of a returning Super Heavy booster.

14 "In 2025 it flew 165 Falcon 9 missions": SpaceX, June 2026 prospectus, reports 165 Falcon 9 launches in 2025, a company-reported vehicle count: https://www.sec.gov/Archives/edgar/data/1181412/000162828026039276/spaceexplorationtechnologi.htm.

14 "single airframes racking up flight counts": SpaceX Form S-1, May 2026; Falcon 9 has demonstrated 34 reflights of a single first-stage booster, with recovery success above 99 percent in 2025.

CHAPTER 2
BREAKING THE LAUNCH CEILING

20 "something on the order of fifty-four thousand dollars": NASA historical cost analyses and Shuttle programme records place the Space Shuttle's average cost to low Earth orbit around $54,500 per kilogram.

20 "an army of technicians months of work": Shuttle programme operational records; refurbishment between flights included thermal-protection-tile inspection, main-engine disassembly, and solid-rocket-booster recovery.

21 "brought a kilogram down toward eight thousand dollars": Compiled from secondary launch-economics analyses, which place ULA's launch vehicles between roughly $6,100 and $15,500 per kilogram.

22 "came to something like two percent": Elon Musk, public statements on SpaceX's founding rationale, widely cited in SpaceX histories; on the idiot index of finished cost to raw-material cost, see Max Olson, "Atoms are Cheap, Process is Pricey" (FutureBlind, February 2026), and Eric Berger, *Liftoff*.

23 "proposed answer was to recover the engines": Spaceflight Now, interview with United Launch Alliance chief Tory Bruno, 14 April 2015, describes SMART reuse of the engine module rather than the whole Vulcan first stage: https://spaceflightnow.com/2015/04/14/ula-chief-explains-reusability-and-innovation-of-new-rocket/.

23 "the night of December 21, 2015": SpaceX historical launch records; Falcon 9's first booster landing, December 2015.

25 "the company facing another failure it could scarcely afford": Elon Musk, 2017 IAC presentation, said the fourth Falcon 1 flight used the last money for that rocket: https://www.spacex.com/media/making_life_multiplanetary_transcript_2017.pdf. The source does not quantify weeks of payroll.

25 "Tom Mueller's propulsion team had brought Merlin": NASA, *Falcon 9 Data Sheet*, records the November 2008 full-duration nine-engine firing at McGregor: https://sma.nasa.gov/LaunchVehicle/assets/spacex-falcon-9-data-sheet.pdf. SpaceX's contemporary account of the earlier July 2008 first nine-engine firing quotes propulsion vice-president Tom Mueller on the team's elation when the engines worked together; reprinted in *Freedom's Phoenix*, 2 August 2008: https://www.freedomsphoenix.com/Article/036561-2008-08-02-private-rocket-to-the-moon-conducts-first-full-9-engine.htm. The scene does not place Mueller at the November stand.

26 "The booster cannot hover": Tim Dodd, interview with Danny Jones, Danny Jones Podcast, episode 370, February 9, 2026, 21:15–22:37, transcript at https://pod.wave.co/podcast/danny-jones-podcast/370-i-found-the-proof-that-ends-the-moon-landing-debate-tim-dodd. Dodd explains why even one Merlin at minimum throttle produces too much thrust for the returning stage to hover, and why an early burn can stop it above the ground and start it rising.

27 "passed ten flights, then fifteen, then twenty": SpaceX Form S-1, May 2026, records 34 reflights of a single first-stage booster; independent launch-tracker data placed the fleet leader, B1067, at 33 flights as of March 2026, with a fleet average of roughly nine to ten flights per booster.

28 "something near twenty-seven hundred dollars a kilogram": SpaceX Form S-1, May 2026, citing NASA historical cost data — roughly $2,700 per kilogram for Falcon 9 (2010) and roughly $1,400 for Falcon Heavy (2018), with Starship aimed at a further

reduction of 99 percent or more against the historical average launch cost; the sub-$200 Starship figure is an aspirational target discussed by SpaceX and outside analysts, not a demonstrated operating cost.

30 "On October 13, 2024, at Boca Chica": SpaceX Starship Integrated Flight Test 5 mission summary, live broadcast, and vehicle specifications page; the first catch of the Super Heavy booster by the launch tower's mechanical arms, nicknamed the chopsticks.

31 "moved that weight off the vehicle and onto the tower": Elon Musk, interviewed by Tim Dodd (Everyday Astronaut) during a walking tour of SpaceX Starbase, Boca Chica, Texas, 2021; on moving mass to the ground side, "that's why we took legs off the booster and just have the tower catch it."

31 "It caught the booster on the first attempt": SpaceX Integrated Flight Test 5 records; the first tower-catch attempt, on the system's fifth full test flight, succeeded.

CHAPTER 3
THE CONSTELLATION

34 "On the twenty-sixth of February he wrote to Musk": Mykhailo Fedorov's 26 February 2022 public request to Elon Musk for Starlink stations was reported contemporaneously by *Ukrainska Pravda*: https://www.pravda.com.ua/eng/news/2022/02/26/7326146/.

34 "the count had climbed into the thousands": CSIS, "Extending the Battlespace to Space," September 2025; more than 50,000 Starlink terminals received in Ukraine over three years of war, connecting command posts, hospitals, and critical infrastructure.

36 "fighting to hold the power grid together": Reuters, October 12, 2022; Ukrainian Vice Prime Minister Mykhailo Fedorov said

Starlink helped restore energy and communications infrastructure in critical areas after missile attacks.

36 "Russian electronic-warfare units": US Department of Defense electronic-warfare official Dave Tremper described SpaceX's software response to Russian jamming in April 2022; *Breaking Defense* contemporaneous report: https://breakingdefense.com/2022/04/spacex-beating-russian-jamming-attack-was-eyewatering-dod-official/.

37 "Starshield for government and national-security use": SpaceX Starshield materials and SpaceX Form S-1, May 2026; Starshield is described as a dedicated government and national-security satellite constellation derived from Starlink.

37 "more than ten thousand working Starlink satellites": Jonathan McDowell, Starlink constellation tracker (2 October 2026), counted 11,142 working Starlink satellites; the text rounds this to more than ten thousand.

37 "near five hundred and fifty kilometers": Official Starlink technical materials; the low-Earth-orbit shell at roughly 550 kilometers, with latency as low as about 25 milliseconds.

38 "turning out about seventy satellites a week": SpaceX, June 2026 prospectus, p. 284, reports that its Redmond, Washington, facility produced an average of approximately 70 Starlink satellites per week from December 2025 through April 2026, a company-reported rate for those months: https://www.sec.gov/Archives/edgar/data/1181412/000162828026041013/japanfwp_06042026.htm.

39 "satellites designed for about five years of service": SpaceX, "Designing and Building Safe, Reliable and Demisable Satellites," distinguishes a five-year satellite design life, active end-of-mission deorbit within weeks, and passive orbital decay within five to six years for non-maneuverable spacecraft: https://www.spacex.com/updates/reusability.

39 "keep launching simply to stand still": Starlink constellation reporting; low-orbit operation requires continuous launch, deorbit, refresh, and capacity thickening as older satellites are retired and replaced.

39 "cost in the hundreds of millions of dollars": Geostationary communications satellite build-and-launch costs in the $200 million to $400 million range are standard satellite-industry estimates; exact totals vary by payload, insurance, and launch arrangement.

40 "Drop that price by an order of magnitude": SpaceX Form S-1, May 2026, citing NASA data; historical average launch cost near $18,500 per kilogram against roughly $2,700 per kilogram on Falcon 9.

40 "twelve million unique service lines": SpaceX, Q2 2026 update, reported twelve million unique service lines at the end of June. Service lines are distinct from subscriber accounts.

41 "a single corridor across Egypt": The International Telecommunication Union estimates submarine cables carry about 99 percent of international internet traffic; TeleGeography and infrastructure analyses place roughly 17 percent of global bandwidth on the corridor through the Suez Canal region; the International Cable Protection Committee and industry reporting record some 100 to 150 cable faults a year, repaired by a fleet of approximately 60 specialized cable ships.

41 "lasers that linked it to its neighbors": Official Starlink technical materials; optical inter-satellite links let the constellation operate as a routed mesh rather than as isolated terminals.

42 "the company answered by darkening the later spacecraft": SpaceX, *Starlink Progress Report 2023*, p. 47, describes first-generation sunshades, darkened surfaces and mirror films, and second-generation dielectric film to reduce observed brightness: https://starlink.com/public-files/starlinkProgressReport_2023.pdf. Its *Brightness Mitigation Best Practices* explains that sun visors were not a viable long-term solution: https://

starlink.com/public-files/BrightnessMitigationBestPracticesSatelliteOperators.pdf.

42 "in partnership with T-Mobile": Starlink, T-Mobile, and FCC materials on the direct-to-cell service for standard phones; the buildout reached more than 650 direct-to-cell satellites within 18 months and more than 400 million people through carrier partnerships including T-Mobile.

44 "The service started with text": Official Starlink, T-Mobile, and FCC materials describe direct-to-cell as supplemental coverage arriving with texting and limited data support first, ahead of broader voice and data parity.

45 "AST SpaceMobile designed large unfolding antennas": AST SpaceMobile, *BlueWalker 3* and its 14 November 2022 deployment announcement, document the 693-square-foot phased array unfolded in orbit for direct-to-phone tests: https://ast-science.com/bluewalker-3/ and https://ast-science.com/693-square-foot-array-on-blue-walker-3-successfully-completed-deployment/.

45 "their cameras looked back at the ship": SpaceX, Starship Flight 12 mission page, 22 May 2026, records deployment of two modified Starlink imaging satellites on a suborbital trajectory: https://www.spacex.com/launches/starship-flight-12. The pictured view is after deployment; this passage does not claim camera formation through atmospheric reentry.

46 "navigate by the constellation itself": Published research investigates Starlink radio signals as an alternative positioning, navigation and timing source: https://arxiv.org/abs/2605.20394. This is research, not an operating Starlink navigation service.

47 "Companies proposed orbital data centers": SpaceX, June 2026 prospectus, pp. 85 and 107, describes proposed AI-compute satellites in Sun-synchronous orbit, explicitly as future infrastructure: https://www.sec.gov/Archives/edgar/data/1181412/000162828026041013/japanfwp_06042026.

htm. Anthropic's 6 May 2026 announcement expresses interest in partnering on orbital compute, not an operating service: https://www.anthropic.com/news/higher-limits-spacex. NASA describes Starship's IFT-3 cryogenic transfer as between tanks within one vehicle, distinct from future ship-to-ship refueling: https://www.nasa.gov/wp-content/uploads/2024/11/lameen-cfmpp-tagged.pdf. Planet imagery is the paragraph's operating example; orbital data centers and refueling depots remain proposals. The AI1 design detail is sourced in Chapter 14.

48 "filed for bankruptcy in 1999": Iridium operated 66 satellites and entered bankruptcy in 1999, within months of beginning commercial service.

CHAPTER 4
THE ENERGY CEILING

53 "the shell of a shuttered appliance factory": The Colossus site occupies the former Electrolux facility on Paul Lowery Road in Memphis; Memphis Light, Gas and Water 2025 xAI update.

54 "a hundred thousand graphics chips": NVIDIA press release, October 28, 2024, on the 100,000-GPU Colossus installation; xAI and NVIDIA later said the Memphis cluster doubled to 200,000 GPUs, and xAI's January 6, 2026 Series E announcement put Colossus I and II together above one million H100-equivalent GPUs, a figure that uses a different measurement unit than raw GPU count.

54 "a running cluster in a matter of months": The first phase was assembled in 122 days, with training beginning 19 days after the first rack was energised; NVIDIA press release, October 28, 2024.

54 "later called the build superhuman": Jensen Huang, speaking on the *Bg2 Pod* with Brad Gerstner and Bill Gurley, October 2024; Huang described the nineteen-day stand-up of the Mem-

phis cluster as superhuman and said there was only one person in the world who could have done it.

55 “synchronized training jobs made the draw rise and fall”: Pacific Northwest National Laboratory, *Data Center EMT Models* (January 2026), pp. 36–37, describes AI-training workload changes and synchronized power variations over short intervals: https://www.energy.gov/sites/default/files/2026-01/Data_Center_EMT_Models.pdf. US Department of Energy, Office of Electricity, “Monitoring Oscillations from Large Data Centers” (May 28, 2026), explains the grid concern: https://www.energy.gov/oe/articles/monitoring-oscillations-large-data-centers. The chapter’s gigawatt is a proposed full-campus scale, not a measured instantaneous swing.

55 “swollen into a backlog measured in years”: PJM Interconnection materials, 2025, reported a total interconnection queue of 195 gigawatts with average wait times of four to five years; Department of Energy materials on AI and data-centre growth tie the expansion to transmission and interconnection backlog.

55 “Their proposed capacity, much of it solar”: US Department of Energy, *Transmission Interconnection Roadmap* (April 2024) and 2024 i2X workshop materials compare approximately 2,600 GW of active queued generation and storage with approximately 1,280 GW of installed US power-plant capacity. Queue entries are proposals, not future production forecasts: https://www.energy.gov/sites/default/files/2024-06/i2X-GTC-CA-Workshop%202024.pdf.

57 “generated several hundred megawatts on private property”: Industry reporting and the regulatory record describe 27 portable gas turbines generating 495 megawatts deployed at the Memphis site before the grid connection was fully ready; the EPA later ruled the turbines had been operated without the required air permits. Memphis Light, Gas and Water 2025 xAI update; EPA enforcement action and industry reporting, 2024–2025.

58 "millions of gallons a day": US Department of Energy, *2024 United States Data Center Energy Usage Report* (Lawrence Berkeley National Laboratory, December 2024), discusses water consumption from data-center cooling and power generation; the chapter describes general demand pressure, not a measured Memphis withdrawal or a verified project delay: https://eta-publications.lbl.gov/sites/default/files/2024-12/lbnl-2024-united-states-data-center-energy-usage-report.pdf.

62 "an unusual annual figure for a small town": Modelled estimates place GPT-3's training around 284,000 kilowatt-hours and GPT-4's closer to 50 gigawatt-hours; these are estimates based on compute requirements and model scale rather than company-disclosed training-energy totals.

63 "began revising their numbers upward": Lawrence Berkeley National Laboratory estimated U.S. data centres used 176 terawatt-hours in 2023 and could reach 325 to 580 terawatt-hours by 2028; Department of Energy materials describe up to 130 gigawatts of new demand by 2028.

63 "a dim light bulb, some twenty watts": The brain's approximately 20-watt power consumption is a well-established figure in the neuroscience literature.

63 "their rated power alone can reach seventy megawatts": NVIDIA, *H100 Tensor Core GPU* specifications, lists up to 700 watts thermal design power for the H100 SXM. One hundred thousand such processors at that rated maximum imply 70 MW for GPUs alone, before networking, cooling and site overhead; this is a calculation, not a metered Colossus total: https://www.nvidia.com/en-us/data-center/h100/.

64 "restart a nuclear reactor that had been shut down": Constellation Energy–Microsoft agreement to restart Three Mile Island Unit 1 as the Crane Clean Energy Center, approximately 835 megawatts, targeted by 2028.

66 "a fifth of all the electricity the country meters": Ireland's Commission for Regulation of Utilities said on December 12, 2025 that data-centre demand had risen from 5 percent of national electricity use in 2015 to 22 percent in 2024 and could reach 31 percent by 2034.

67 "moving assembly line he had perfected at Highland Park": David A. Hounshell, *From the American System to Mass Production, 1800–1932* (Johns Hopkins University Press, 1984), documents the Highland Park line and the reduction of Model T assembly time from roughly 12 hours to 93 minutes.

69 "a hundred thousand people on a single site": The Henry Ford institutional history: the Rouge plant employed more than 100,000 workers and produced 4,000 vehicles per day at peak, while still relying on approximately 6,000 external suppliers.

69 "a town he called Fordlandia": The Henry Ford, archival history of Henry Ford's Amazon rubber plantation and company town at Fordlandia: https://www.thehenryford.org/collections-and-research/digital-collections/.

CHAPTER 5
SOLAR, STORAGE, AND THE MEGAPACK

74 "Hornsdale sent about 7.3 megawatts into the network": Neoen, *Hornsdale Power Reserve*, presented through ARENA (2019), pp. 9–10, records the 14 December 2017 Loy Yang trip, 560 MW loss and a battery response in milliseconds: https://arena.gov.au/assets/2019/02/hornsdale-power-reserve.pdf. AEMO, *Initial Operation of the Hornsdale Power Reserve* (April 2018), pp. 4, 7, records the battery's 100 MW rating and rapid contingency response, while warning full service delivery had not yet been demonstrated: https://www.aemo.com.au/-/media/Files/Media_Centre/2018/Initial-operation-of-the-Hornsdale-Power-Reserve.pdf. The event trace and approximately 7.3 MW dispatch are also described in RenewEconomy, 19 De-

cember 2017: https://reneweconomy.com.au/tesla-big-battery-outsmarts-lumbering-coal-units-after-loy-yang-trips-70003/. These records do not establish an averted cascade.

75 "Hornsdale brought another response alongside inertia": AEMO, *Initial Operation of the Hornsdale Power Reserve* (April 2018), pp. 5–7, describes fast regulation and contingency-frequency service from the battery while distinguishing its enabled controls from the wider grid response: https://www.aemo.com.au/-/media/Files/Media_Centre/2018/Initial-operation-of-the-Hornsdale-Power-Reserve.pdf. Rotational inertia begins resisting a frequency change physically; the battery supplies controlled power after sensing it.

75 "earning a steady income in the market for grid stability": AEMO, *Initial Operation of the Hornsdale Power Reserve* (April 2018), pp. 4–8, documents rapid and precise frequency-control service; Neoen, *Hornsdale Power Reserve* (ARENA, 2019), pp. 8–9, describes reductions in regulation-service costs: https://www.aemo.com.au/-/media/Files/Media_Centre/2018/Initial-operation-of-the-Hornsdale-Power-Reserve.pdf and https://arena.gov.au/assets/2019/02/hornsdale-power-reserve.pdf.

76 "it had dropped under a hundred and fifty": BloombergNEF annual battery price survey and IEA data; average pack prices fell from approximately $1,400 per kilowatt-hour in 2010 to below $140 by 2023, with global averages near $108 per kilowatt-hour by 2024 and Chinese manufacturers lower still.

76 "no longer needed scarce cobalt": International Energy Agency, *Batteries and Secure Energy Transitions* (2024), describes lithium iron phosphate chemistry as using iron and phosphate instead of nickel and cobalt: https://www.iea.org/reports/batteries-and-secure-energy-transitions.

77 "falling by something like ninety percent over the decade": IRENA, *Renewable Power Generation Costs*; the levelized cost of utility-scale solar electricity fell from $0.417 per kilowatt-

hour in 2010 to $0.043 in 2024, a decline of roughly ninety percent.

80 "wired a Porsche 944 to run on batteries": JB Straubel, *Ride the Lightning* podcast, Episode 501, interviewed by Ryan McCaffrey, March 2025; Straubel describes converting a Porsche 944 to battery power in the garage of his college rental house, writing the control software himself, mostly in assembly language, for a car with about twenty miles of range. In the same March 2025 interview Straubel recalls watching power lines from a car as a child, fascinated by storing, moving and transforming electricity.

80 "first meeting Musk after a talk": JB Straubel, *Ride the Lightning*, episode 501 (9 March 2025), recalls first meeting Musk after a Stanford talk on early SpaceX; he raised electric airplanes, then they discussed lithium-ion batteries and electric cars: https://www.youtube.com/watch?v=zLgzpoSllgM.

80 "He approached Musk with an electric-airplane idea": JB Straubel, *Ride the Lightning*, episode 501 (9 March 2025), recalls approaching Musk with an electric-airplane idea after a Stanford talk on early SpaceX, then discussing lithium-ion batteries and electric cars: https://www.youtube.com/watch?v=zLgzpoSllgM.

81 "the end of that approach, and it was supply": JB Straubel, *Ride the Lightning* podcast, Episode 501, March 2025; Straubel recalls that in trying to build an electric vehicle at the time, the batteries were the heart of the problem that needed major improvement to be practical.

81 "work on that capacity before the Model 3": JB Straubel, *Ride the Lightning* podcast, Episode 501, March 2025; Straubel describes the first Gigafactory as a giant supply chain problem rather than a glamorous project, the repeated trips to Japan with Musk to win Panasonic's commitment, and the team that found the Nevada site, built the factory, and stood it up into production.

82 "founded a second company, Redwood Materials": JB Straubel founded Redwood Materials, a battery recycling and remanufacturing company, in 2017, two years before leaving Tesla; *Ride the Lightning* podcast, Episode 501, March 2025.

82 "a mine that needed no digging": JB Straubel, *Ride the Lightning* podcast, Episode 501, March 2025; Straubel describes Redwood as "a second almost unmanufacturing plant" and lays out a closed-loop vision in which, at scale, very little new mining is needed at all.

83 "nearly four megawatt-hours of storage": Tesla Megapack product pages; each unit holds approximately 3.9 megawatt-hours of storage, delivers 1.9 megawatts of power, and operates at up to 93 percent round-trip efficiency.

83 "tens of gigawatt-hours of these things a year": Tesla investor materials and the Q4 FY2025 update; the Lathrop Megafactory carries a nameplate capacity of 40 gigawatt-hours a year, with the Shanghai Megafactory adding roughly the same from early 2025. These are company-stated capacity figures rather than independently audited throughput.

85 "Mike Snyder, who runs Tesla": American Clean Power Association, Michael Snyder board biography (2026), identifies him as Tesla's Vice President of Energy and Charging and says he leads that organization, including Megapack, Powerwall, solar and Supercharging: https://cleanpower.org/about/team/michael-snyder/.

85 "putting cabinets beside the gas turbines": Memphis Light, Gas and Water 2025 xAI update; Greater Memphis Chamber report, May 7, 2025; the xAI Memphis data center paired roughly 150 megawatts of grid power with 150 megawatts of Tesla Megapack battery backup alongside temporary gas turbine generation.

87 "Tesla deployed 13.5 gigawatt-hours of storage": Tesla, *Q2 2026 Update*, pp. 4–5, reports 13.5 GWh of energy-storage deployment during the three-month second quarter of 2026; this is a company-reported quarterly deployment figure, not annual

output or factory capacity: https://ir.tesla.com/_flysystem/s3/sec/000162828026049213/tsla-20260722-gen.pdf.

88 "one of the major sources feeding its grid": California Independent System Operator, battery-storage and daily supply reporting, documents the scale and timing of California battery dispatch; the passage does not assert the largest resource at a particular interval: https://www.caiso.com/todays-outlook/supply.

CHAPTER 6
THE FACTORY AS TECHNOLOGY

91 "on the order of six thousand tonnes": Idra Group, *Giga Press* product materials, specifies six-thousand-tonne-class high-pressure die-casting machines: https://idragroup.com/index.php/en/gigapress.

93 "A die at this scale can cost millions": JST Consultancy, "Heat treatment of GIGA dies for GIGA Casting," estimates €2–4 million for a large die, illustrating the capital scale rather than a disclosed Tesla purchase price: https://www.jstconsultancy.cz/en/heat-treatment-of-giga-dies-for-giga-casting/. The month-scale lead time reflects large-tooling design and manufacture, not a verified clock for the particular die described here.

96 "without a later heat treatment that could warp": Tesla, *Battery Day 2020*, described its new high-strength aluminium casting alloy as not requiring heat treatment or coating; avoiding the subsequent bake matters for a very large casting, but the source does not promise zero production defects: https://www.tesla.com/2020shareholdermeeting.

98 "The roughly seventy stamped parts it replaced": Tesla, 2020 Battery Day and manufacturing presentations, described a single rear underbody casting replacing roughly seventy stamped pieces; this is a part count, not a count of suppliers or purchase orders: https://www.tesla.com/2020shareholdermeeting.

98 "a two-million-dollar robot cell": Elon Musk, interviewed by Tim Dodd (Everyday Astronaut) during a walking tour of SpaceX Starbase, Boca Chica, Texas, 2021; Musk recounts the fiberglass-mat episode on the Model 3 battery-pack line in detail, including the contradictory battery-safety and noise-and-vibration explanations, the microphone comparison test, and the deletion of the mats along with the two-million-dollar robot cell.

98 "question the requirement, delete the part": The five-step process — make requirements less dumb, delete the part or process, simplify, accelerate cycle time, and only then automate — as laid out by Elon Musk during the same Starbase walking tour with Tim Dodd, 2021; Musk adds that he has personally run the steps backward "multiple times."

99 "Excessive automation at Tesla was a mistake": Elon Musk, posting publicly on April 13, 2018: "Excessive automation at Tesla was a mistake. To be precise, my mistake. Humans are underrated."

100 "Musk had made the production floor a center of his attention": Elon Musk, interviewed by Tim Dodd during a Starbase tour (30 July 2021), argued that manufacturing had received too little attention relative to design, explaining how much work went into a production system: https://everydayastronaut.com/starbase-tour-and-interview-with-elon-musk/.

100 "He spent his nights on the production floor": Elon Musk, Starbase walking tour with Tim Dodd (Everyday Astronaut), 2021; Musk describes having been "basically living on the battery pack production line" during the Model 3 ramp.

100 "Sandy Munro had spent his career taking cars apart": Munro & Associates, *About*, describes Sandy Munro's automotive manufacturing career and the firm's teardown and benchmarking practice: https://www.munrocommercial.com/about/. The firm's 20 April 2021 announcement recounts its Model Y teardown: https://www.prnewswire.com/news-releases/munro-

associates-inc-expands-staff-opens-new-facility-in-response-to-record-setting-growth-301272550.html.

102 "In 2023 Tesla showed where the casting was heading": Tesla, *Investor Day 2023*, presented the proposed unboxed manufacturing process and its parallel subassemblies: https://www.tesla.com/investor-day. The chapter describes the process prospectively.

102 "Henry Ford had set running at Highland Park in 1913": David A. Hounshell, *From the American System to Mass Production, 1800-1932* (Johns Hopkins University Press, 1984), and Ford Motor Company institutional histories on the moving assembly line and the reduction of Model T assembly time from roughly 12 hours to 93 minutes.

103 "Lars Moravy, who ran vehicle engineering at Tesla": Tesla, *Investor Day 2023*, identifies Lars Moravy as Vice President of Vehicle Engineering and presents his proposed unboxed production process: https://www.tesla.com/investor-day.

CHAPTER 7
THE LABOR CEILING

107 "Give a Tesla Optimus the eighteen-meter walk": This opening is a constructed factory scenario, not a report of a documented Fremont night shift or an established Optimus pack-delivery deployment. The eighteen-meter route, clock times, cart hand-offs and crew details illustrate the work a humanoid would have to learn. Public Optimus demonstrations and Tesla's stated training plans supply the technological setting; they do not verify these particular events.

109 "the hand closed on a battery cell": Tesla's publicly released Optimus demonstration footage shows handling and seating battery cells in a tray. This is a demonstration of the motion, not evidence that the constructed pack-route opening is an operating installation. The hand's tactile/control explanation concerns its design and the engineering problem.

110 "the hardest tasks to put into a machine": The observation that sensorimotor tasks easy for humans are hard for machines, while abstract reasoning runs the other way, is Moravec's paradox; Hans Moravec, *Mind Children: The Future of Robot and Human Intelligence*, Harvard University Press, 1988.

112 "five million industrial robots were already operating worldwide": International Federation of Robotics, *World Robotics 2026*, counted about five million industrial robots operating worldwide in 2025.

112 "owned since 2016 by a Chinese appliance maker": KUKA, 2016 investor agreement with Midea, documents the acquisition process and continuing company identity: https://www.kuka.com/en-de/company/press/news/2016/06/investor-agreement-midea.

115 "a price below thirty thousand dollars": Tesla's stated target is a future Optimus price below $30,000 at scale, not a current market price.

116 "first Optimus builds to an Academy for training": Tesla, Q2 2026 update, described an Optimus Academy training path for initial builds.

117 "takes totes from mobile robots and puts them on a conveyor": Agility Robotics and GXO, Flowery Branch deployment materials, describe Digit moving totes from mobile robots to a conveyor; Agility reported more than 100,000 cumulative totes.

119 "Ashok Elluswamy's team carries the AI work for Optimus": Tesla, *Investor Day 2023*, presented Ashok Elluswamy in both its autonomy and Optimus sections: https://www.tesla.com/investor-day. Tesla's *AI & Robotics* page places vehicle autonomy and Optimus within its AI program: https://www.tesla.com/AI. Bloomberg reported his assumption of Optimus responsibility after Milan Kovac's departure in June 2025: https://www.bloomberg.com/news/articles/2025-06-06/tesla-s-leader-of-optimus-humanoid-robot-program-leaves-company.

CHAPTER 8
THE UNPROTECTED LEFT

123 "Picture a weekday evening in San Francisco": The Mission District ride is constructed to illustrate perception, planning and supervised driving. Its 6:47 time, pedestrian, cyclist, passenger and exact gap are not a documented trip. Tesla's Q2 2026 update, p.9, labels the San Francisco Bay Area operation Safety Driver and describes FSD (Supervised): https://ir.tesla.com/_flysystem/s3/sec/000162828026049213/tsla-20260722-gen.pdf. The scene retains that driver; it does not portray an authorized driverless Cybercab service in San Francisco.

124 "through cameras alone. Eight of them": Andrej Karpathy, interview with Lex Fridman, *Lex Fridman Podcast*, episode 333, 2022; Karpathy describes the network reading all eight cameras at once and makes the engineering case for vision-only driving — cameras as the highest-bandwidth sensor, with lidar and other added sensors carrying a full organizational cost beyond their purchase price.

125 "several outcomes remain possible": Tesla's *Q4 2025 Update* (January 28, 2026) describes FSD (Supervised) as using an end-to-end foundation model: https://ir.tesla.com/_flysystem/s3/sec/000162828026003837/tsla-20260128-gen.pdf. The pedestrian, cyclist and changing gap in this constructed Mission District ride illustrate decisions the car has to make; they are not a description of a disclosed branch-and-cost planning algorithm.

127 "It could learn to move a battery pack": Tesla, *Q2 2026 Update* (July 22, 2026), reports initial Optimus production builds entering its Academy training program: https://ir.tesla.com/_flysystem/s3/sec/000162828026049213/tsla-20260722-gen.pdf. The battery-pack route is a constructed example of a task such a robot might learn; the update does not report this exact operation.

129 "in the robot business and the AI business at once": Andrej Karpathy, interview with Lex Fridman, *Lex Fridman Podcast*, episode 333, 2022; Karpathy describes Optimus reusing the Autopilot stack — "this robot currently thinks it's a car" — with the vision system, data engine, and training infrastructure carrying over from car to robot through retraining rather than redesign.

129 "He called that Software 2.0": Andrej Karpathy, interview with Lex Fridman, *Lex Fridman Podcast*, episode 333, November 2022; Karpathy describes Software 2.0 as a compilation process in which datasets, objectives, and architecture compile into neural-network weights, with training producing the deployable binary.

132 "numbered in the millions of vehicles": Andrej Karpathy, interview with Lex Fridman, *Lex Fridman Podcast*, episode 333, 2022; Karpathy frames the decisive question in autonomy as "do you have the fleet or not," with Tesla's million-plus customer cars serving as the data-collection instrument as of 2022.

132 "the relevant seconds of camera footage uploaded": Andrej Karpathy, interview with Lex Fridman, *Lex Fridman Podcast*, episode 333, 2022; Karpathy describes the data engine's feedback loop — deploy, gather edge cases, mine, label, retrain — by which the fleet's hard moments become the next version's training set.

132 "a single network trained from end to end": Andrej Karpathy, interview with Lex Fridman, *Lex Fridman Podcast*, episode 333, 2022; Karpathy recounts the progressive replacement of handwritten code by neural networks inside Autopilot — "we want the neural nets to write the algorithm" — from single-image predictions to networks spanning all eight cameras and time.

134 "the fleet had captured a rare event": Andrej Karpathy, interview with Lex Fridman, *Lex Fridman Podcast*, episode 333, 2022; Karpathy argues the lidar-versus-camera debate misses the structural point — "the actual debate should be about do

you have the fleet or not" — contrasting test fleets numbered in the hundreds with millions of customer-owned cars gathering data at scale.

134 "rebuild that moment in simulation": Tesla's Full Self-Driving materials describe fleet data and simulation as training inputs. This reconstructed intersection event illustrates that method.

134 "what the engineers call shadow mode": Tesla, *Autonomy Investor Day* (2019), described fleet learning and shadow-mode evaluation as part of the Autopilot development process: https://www.tesla.com/autonomyday.

135 "Miles per intervention, the average distance": Andrej Karpathy, interview with Lex Fridman, *Lex Fridman Podcast*, episode 333, 2022; Karpathy recalls early Autopilot needing three or four interventions on a Palo Alto–San Francisco drive, with capability gains arriving as a staircase of deploy-and-retrain cycles over the following five years.

CHAPTER 9
TWO OCEANS OF DATA

139 "It was fourteen minutes past two in the morning": Illustrative constructed scene. Tesla, Privacy Notice (https://www.tesla.com/legal/privacy), describes opt-in Fleet Learning camera recordings; Tesla, Full Self-Driving (Supervised) (https://www.tesla.com/fsd), describes training from fleet driving data. Neither establishes this Tennessee deer event, clock time, driver, packet size or upload.

141 "To the people who run the pipeline that": Tesla FSD evidence dashboard (https://www.tesla.com/fsd-evidence-dashboard), Eight exterior cameras, IMUs, GPS and in-vehicle AI compute

141 "the fleet numbered several million cars": Tesla Q4 and FY2025 Update; Tesla reported deliveries of 1,789,226 vehicles in 2024 and 1,636,423 in 2025, consistent with a cumulative global fleet in the several millions.

141 "carries eight cameras looking out in every direction": Andrej Karpathy, *Lex Fridman Podcast*, episode 333, 2022; Karpathy describes Tesla's perception stack running on eight cameras over time rather than a single camera.

143 "rooms full of people drawing boxes": Andrej Karpathy, *Lex Fridman Podcast*, episode 333, 2022; Karpathy describes growing Tesla's annotation team from zero to roughly a thousand people, with labeling tasks deliberately split between human annotators and the automated offline tracker.

143 "The pipeline's answer was to automate the labeling": Andrej Karpathy, *Lex Fridman Podcast*, episode 333, 2022; Karpathy describes the offline tracker, an automatic reconstruction process that runs large neural networks against fleet video after the fact, with no in-car time constraints, to recover three-dimensional ground truth.

144 "knew its own gaps and reached out to fill them": Andrej Karpathy, *Lex Fridman Podcast*, episode 333, 2022; Karpathy's account of the data engine, in which deployed networks surface uncertain and underrepresented scenarios, the fleet is mined for matching examples, and the training set is padded out where it is thin.

145 "go out over the air to the fleet": Tesla official Full Self-Driving materials describe a camera-first system trained on fleet data and simulation and deployed back to vehicles through over-the-air software updates.

146 "There was a question of whose data this": Tesla privacy notice (https://www.tesla.com/legal/privacy), AI Data / Fleet Learning camera recordings: opt-in; no continuous capture

146 "In 2022 Musk had bought Twitter": Transaction records; Musk completed the acquisition of Twitter in October 2022.

147 "that combined entity was folded into SpaceX": xAI acquired X at a $113 billion combined valuation in March 2025, per Reuters and xAI announcements; xAI's official announcement

of February 2, 2026 and Reuters reporting the same day record the combination with SpaceX.

147 "By the time X entered SpaceX": Epoch AI, forecast of available public human text (https://arxiv.org/abs/2211.04325), projects that public human-text stock could be fully utilized during 2026–2032. This is a forecast, not evidence that text had already been exhausted when X entered SpaceX.

148 "three hundred and fifty million posts a day": SpaceX Form S-1, 2026; the filing reports approximately 350 million daily posts on X.

149 "Tesla held the road footage and SpaceX held": SpaceX Form S-1 (https://www.sec.gov/Archives/edgar/data/1181412/000162828026036936/spaceexplorationtechnologi.htm) dates its acquisition of xAI to 2 February 2026, after xAI acquired X. Tesla, Privacy Notice (https://www.tesla.com/legal/privacy), describes Tesla-held fleet-learning recordings. The speaking-car combination is imagined, not a disclosed shared training corpus.

150 "Tesla used simulation in its driving work": Tesla, Q3 2021 Update, Core Technology (https://ir.tesla.com/_flysystem/s3/sec/000156459021051307/tsla-8k_20211020-gen.pdf), names simulation alongside fleet collection, labeling and network training for FSD. The varied intersection illustrates the method, not a disclosed xAI simulation corpus.

150 "Manufacturing data carried its own danger, and the": Nature (2024) (https://www.nature.com/articles/s41586-024-07566-y), Recursive training on model-generated data and model collapse

152 "Tesla's fleet poured in the physical world": Tesla Full Self-Driving Vehicle Safety Report (https://www.tesla.com/fsd/safety) describes fleet telemetry categories. This source supports the road-data stream, not the X platform or a merged training library.

CHAPTER 10
CHIPS AS CHOKE POINT

155 "Roughly a hundred and fifty kilometers off the": CIA World Factbook / Britannica (https://www.cia.gov/the-world-factbook/about/archives/2021/countries/taiwan/), Taiwan and Belgium geography, population and strait distance; CIA World Factbook / Britannica (https://www.cia.gov/the-world-factbook/about/archives/2021/countries/belgium/), Taiwan and Belgium geography, population and strait distance; CIA World Factbook / Britannica (https://www.britannica.com/place/Taiwan-Strait), Taiwan and Belgium geography, population and strait distance

156 "ninety percent of the world's most advanced logic chips": Semiconductor industry analyses and TrendForce-style market-share work commonly place TSMC at roughly 90 percent of the world's most advanced chipmaking capacity, with TSMC, Samsung, and Intel the only firms fabricating at 5 nanometres and below by 2025.

156 "The wager was TSMC": TSMC was founded in 1987 by Morris Chang, establishing the dedicated-foundry model that separated chip design from chip fabrication; see Chris Miller, Chip War: The Fight for the World's Most Critical Technology (Scribner, 2022).

157 "In 2014 IBM agreed to pay GlobalFoundries roughly": IBM 2014 GlobalFoundries announcement (https://www.ibm.com/investor/att/pdf/IBM-To-Divest-Microelectronics-Business-To-GLOBALFOUNDRIES.pdf), 2014 agreement, $1.5 billion consideration, target closing in 2015

157 "IBM agreed to pay GlobalFoundries roughly a billion": IBM, "IBM to Divest Microelectronics Business to GLOBALFOUNDRIES," 20 October 2014 (https://www.ibm.com/investor/att/pdf/IBM-To-Divest-Microelectronics-Business-To-GLOBALFOUNDRIES.pdf), pp.1–2: IBM agreed to pay $1.5 billion over three years. IBM 3Q 2015 Form 10-

Q (https://www.sec.gov/Archives/edgar/data/51143/000005114315000009/ibm15q3_10q.htm), Divestitures: transaction closed 1 July 2015.

157 "costs on the order of a few hundred million dollars": ASML corporate disclosures and industry reporting commonly put EUV system pricing around $200 million per machine, with High-NA EUV around $350 million; the same reporting documents the machines' part counts and transport and installation logistics.

158 "This dependence was thought to give Taiwan": Council on Foreign Relations (https://www.cfr.org/articles/war-over-worlds-most-critical-technology-conversation-chris-miller), Taiwan chip concentration and the debated silicon-shield argument; Council on Foreign Relations (https://www.cfr.org/articles/will-chinas-reliance-taiwanese-chips-prevent-war), Taiwan chip concentration and the debated silicon-shield argument

159 "C.C. Wei, an engineer who had risen through": TSMC executive biography (https://www.tsmc.com/english/aboutTSMC/executives), C.C. Wei chairman and CEO since June 2024; TSMC 2024 annual report (https://investor.tsmc.com/static/annualReports/2024/english/index.html), Taiwan N2 development alongside overseas Arizona and Japan expansion; TSMC 2025 earnings transcript (https://investor.tsmc.com/chinese/encrypt/files/encrypt_file/reports/2025-04/7630274eecc1197a4e3ea6a415f44a47204fe10a/TSMC%201Q25%20Transcript.pdf), Overseas decisions based on customer needs and support; advanced expansion includes Arizona

159 "The obvious answer to a concentration that dangerous": TSMC 2025 AGM minutes (https://investor.tsmc.com/sites/ir/shareholders-meeting/2025-06-03/2025AGM_Minutes_wmn_0.pdf), Arizona Fab 1 N4 high-volume production in Q4 2024, earlier than scheduled

162 "a local manufacturing ecosystem has to be cultivated": TSMC Arizona announcement (https://pr.tsmc.com/system/files/newspdf/attachment/706c28f159690dcd2180778a2ff26827d0d898e0/TSMC%20Arizona%20Announcements%20Press%20Release%20%28with%20event%20and%20quotes%29_%28E%29_final_wmn.pdf) describes building a local semiconductor ecosystem with partners. TSMC 2025 AGM minutes (https://investor.tsmc.com/sites/ir/shareholders-meeting/2025-06-03/2025AGM_Minutes_wmn_0.pdf) date Fab 1 N4 high-volume production to Q4 2024, earlier than its revised schedule. The mature ecosystem claim extends beyond first production; no exact decade is asserted.

164 "eleven million vehicles went unbuilt": Industry analyses and automotive trade publications commonly put the 2021 semiconductor-shortage hit at roughly 11 million vehicles of lost production and more than $200 billion in lost revenue.

164 "nineteen new controller variants": Elon Musk public statements and Tesla earnings calls; Tesla developed 19 new controller variants to adapt to alternative chips during the 2020-2021 semiconductor shortage.

164 "While the public argument about Tesla and chips": Tesla Q4/FY2024 update (https://digitalassets.tesla.com/tesla-contents/image/upload/IR/TSLA-Q4-2024-Update.pdf), Q4 completion of Cortex ~50k H100 at Giga Texas

164 "The cluster has a name, Cortex": Tesla Q4 and FY2025 Update and Tesla investor materials describe the Cortex AI training cluster at Giga Texas, by then exceeding 100,000 H100-equivalent GPUs; the figure is company-described infrastructure rather than an independently audited hardware inventory.

165 "A cluster like Cortex is easier to name": Tesla Q4/FY2024 update (https://digitalassets.tesla.com/tesla-contents/image/upload/IR/TSLA-Q4-2024-Update.pdf), Cortex contained approximately 50,000 H100s; NVIDIA H100 specifications (https://www.nvidia.com/en-in/data-center/h100/),

H100 SXM up to 700 W TDP; illustrative 50,000 x 0.7 kW x 8,760 h x $0.10/kWh = $30.66m annual GPU-only full-load electricity, before overhead

165 "The H100 was a single graphics processor made": NVIDIA FY2024 annual report (https://www.sec.gov/Archives/edgar/data/1045810/000104581024000029/nvda-20240128.htm), Data-center demand and manufacturing supply constraints

166 "cost into the tens of thousands of dollars": Outside teardown models, industry reporting, and NVIDIA financial results commonly placed H100 manufacturing cost near $3,320 and average selling price near $28,000; NVIDIA does not publish unit-level economics.

167 "Tesla reached that conclusion as part of a": Google Cloud TPU announcement 2017 (https://blog.google/innovation-and-ai/infrastructure-and-cloud/google-cloud/google-cloud-offer-tpus-machine-learning/), Google TPUs used for training and serving production machine-learning models

167 "The largest cloud providers had each built their own": Amazon's February 5, 2026 results said its Trainium and Graviton chips had reached a combined annual revenue run rate above $10 billion.

167 "The attempt to make those chips had a": TechCrunch, Dojo team breakup (7 August 2025) (https://techcrunch.com/2025/08/07/tesla-shuts-down-dojo-the-ai-training-supercomputer-that-musk-said-would-be-key-to-full-self-driving/), Breakup of the original Dojo team and 2025 shift toward AI6; Tesla Q1 2026 update (https://ir.tesla.com/_flysystem/s3/sec/000162828026026551/tsla-20260422-gen.pdf), Continuing custom silicon development with Dojo 3

168 "Musk's framing at the time was that every": Walter Isaacson, Elon Musk (2023), Manufacturing chapter (https://www.simonandschuster.net/books/Elon-Musk/

Walter-Isaacson/9781982181307), 2018 Jupiter/walk-to-red scene, roughly one hundred decisions and the full quoted remark; this is an analogy for the later Dojo decision; Musk August 2025 statement reported by TechCrunch (https://techcrunch.com/2025/08/11/elon-musk-confirms-shutdown-of-tesla-dojo-an-evolutionary-dead-end/), Musk said Dojo 2 was an evolutionary dead end and paths converged to AI6 at the time; later Dojo 3 continued

168 "AI6 was the chip Tesla actually wanted, and": TechCrunch (2025) (https://techcrunch.com/2025/07/28/tesla-signs-16-5b-deal-with-samsung-to-make-ai-chips/), July 2025 Tesla-Samsung $16.5 billion AI-chip agreement

169 "The organization that has to turn the AI6 design": DCD April 2026 Nivarti report (https://www.datacenterdynamics.com/en/news/anant-nivarti-rejoins-tesla-to-lead-silicon-engineering-including-ai6-and-dojo-3/), Nivarti silicon-engineering lead including AI6 and Dojo 3

170 "The reflex that would eventually produce Terafab had": Walter Isaacson, Elon Musk (2023), Manufacturing chapter (https://www.simonandschuster.net/books/Elon-Musk/Walter-Isaacson/9781982181307), Nevada battery Gigafactory plan, staged bulldozers near Reno, Tsuga dinner and Panasonic financing about $2 billion of the projected $5 billion cost

171 "The Terafab idea reached below chip design": SpaceX 2026 S-1 (https://www.sec.gov/Archives/edgar/data/1181412/000162828026036936/spaceexplorationtechnologi.htm), Terafab general framework, unsettled commercial terms and expected Intel role

171 "Intel held the process knowledge and packaging equipment": SpaceX Form S-1, May 2026; the filing describes Intel as having "the ability to design, fabricate, and package ultra-high-performance chips at scale," and describes a Terafab strategy of verti-

cally integrating mask design, logic and memory fabrication, and advanced packaging in a single closed-loop plant.

172 "pouring money into rebuilding American fabrication": The CHIPS and Science Act, signed into law August 9, 2022, authorised approximately $52.7 billion for semiconductor manufacturing, research, and workforce development, including $39 billion in manufacturing incentives; Intel's award was $7.865 billion.

172 "The discipline Musk had drilled into Tesla and": Everyday Astronaut Starbase factory tour, Part 1 (https://www.youtube.com/watch?v=t705r8ICkRw), Musk five-step process and adding-things-back-in-ten-percent remark

172 "you're clearly not deleting enough": Elon Musk, Starbase factory tour with Tim Dodd (Everyday Astronaut), Boca Chica, Texas, 2021; Musk laid out the five-step process on camera and said, in slightly different phrasings each time, "If you're not adding things back in 10% of the time, you're clearly not deleting enough."

173 "one terawatt of computing hardware a year": SpaceX Form S-1, May 2026; the filing describes the Terafab initiative with Tesla to build a manufacturing facility capable of producing 1 terawatt per year of compute hardware, with Intel joining the project in early April 2026.

174 "The Apollo case is the one that rhymes": NASA historical study (https://www.nasa.gov/wp-content/uploads/2016/01/historical-studies-societal-impact-spaceflight-ebook_tagged.pdf), Apollo purchased over one million integrated circuits from 1962 to 1967, stimulating scale and price decline

174 "In the same window, at Bell Labs, Mervin": Jon Gertner, The Idea Factory (2012) (https://www.penguinrandomhouse.com/books/303275/the-idea-factory-by-jon-gertner/), Kelly and Bell Labs institutional history; Western Electric company history (https://www.westernelectric.com/company-history),

Western Electric was the Bell System manufacturing arm using Bell Labs engineering drawings

174 "Kelly, who ran the labs": The account of Mervin Kelly and Bell Labs draws on Jon Gertner, *The Idea Factory: Bell Labs and the Great Age of American Innovation* (Penguin Press, 2012).

CHAPTER 11
ONE DAY, FOUR SITES

177 "the day that brings their jobs together is imagined": The four-site October day is an illustrative scene. Its people, exact clock times, robot trial, freight route, tests and cross-site handoffs are constructed. The sites and underlying machines have separate public evidence in the adjacent chapter notes.

178 "two Optimus units are carrying empty parts trays": Tesla, Q4 2025 earnings call, described Optimus Gen 3 as intended for later mass production. The two-unit tray trial belongs to this chapter's constructed day.

180 "three Tesla Semis are east of Reno": Tesla, Q4 2025 update, describes the Semi program. The three-truck route and cargo belong to this chapter's constructed day.

185 "the Cortex 2 cluster finishes an inference run": Tesla Q2 2026 update (https://ir.tesla.com/_flysystem/s3/sec/000162828026049213/tsla-20260722-gen.pdf), Supporting Infrastructure table labels Cortex 2 production and >115 MW installed; October-day inference run and electrical swing are constructed scene elements, not claims in Tesla's Q2 update.

186 "SpaceX proposed carrying computing into orbit": SpaceX's January 2026 FCC application described orbital computers in shells, powered by solar arrays and joined to an existing space network. SpaceX Form S-1, May 2026, as amended, also describes space AI infrastructure as a company proposal rather than operating capacity.

186 "A selected sun-synchronous orbit can keep a solar array": SpaceX Form S-1, May 2026, as amended; the filing states that space-based solar arrays can generate more than five times the energy per unit area of terrestrial solar, owing to continuous illumination, lack of atmospheric interference, and optimal orientation.

186 "Optical links would connect the compute spacecraft": SpaceX January 2026 FCC application describes optical inter-satellite connections and routing through Starlink to authorized ground stations. This is a proposed route, not an operating orbital-compute network.

187 "Google's small orbital-compute prototype reached orbit": Google, 'Project Suncatcher prototype in orbit' (1 October 2026) says the Planet-built prototype launched on SpaceX Transporter-18 and contact was confirmed. The mission was to gather TPU radiation, thermal and physical-stress data. It establishes a test spacecraft, not commercial orbital-compute service or a SpaceX compute-rental contract.

CHAPTER 12
WHERE THE LAYERS MEET

192 "The teams put the design through hundreds of test and validation cycles": Volkswagen Group, winter-testing update (March 2026) describes reference vehicles from Volkswagen, Audi and Scout and reports hundreds of test and validation cycles. The count is the company's report, not an independently audited tally.

192 "up to $5.8 billion in Rivian and the venture": Volkswagen Group, joint-venture announcement (2024) describes an investment and venture commitment of up to $5.8 billion.

194 "Production of that electrical design began in January 2026": Volkswagen Group China, CEA production announcement (2026).

194 "send its software specialists into the joint venture": Volkswagen Group, winter-testing update (March 2026) describes a planned specialist-training program.

195 "Starliner carried Butch Wilmore and Sunita Williams": NASA, Boeing Crew Flight Test return decision (August 2024) documents the flight and crew.

196 "When Starliner could not carry its first crew home": NASA, Crew-9 return account (March 2025) documents the separate return route.

197 "In 1981 IBM assembled its personal computer": IBM, personal-computer history describes the 1981 PC's open architecture and externally sourced components.

197 "Morris Chang founded TSMC in 1987": Morris Chang founded Taiwan Semiconductor Manufacturing Company in 1987, establishing the dedicated-foundry model that separated chip design from chip fabrication; see Chris Miller, Chip War: The Fight for the World's Most Critical Technology (Scribner, 2022).

199 "BYD began as a battery company in 1995": BYD, company history dates its battery-company founding to 1995.

200 "Cells had to become packs": BYD corporate disclosures and company materials document the company's breadth across batteries, motors, power electronics, storage, and vehicle manufacturing.

200 "The Blade Battery is set into the vehicle structure": BYD, Cell-to-Body brochure describes Blade Battery integration with vehicle structure.

201 "Its hard-carbon process called for moisture control": CATL, sodium-ion process announcement (April 2026) describes moisture control in hard-carbon production and process changes intended to improve output consistency. These are CATL's stated process aims, not independently verified yield results.

201 "Transmission lines carry power across provinces": State Grid project completion report (2025) describes a long-distance high-voltage line.

203 "Ore, glass, steel and rubber entered a vast complex": The Henry Ford institutional history; David A. Hounshell, *From the American System to Mass Production, 1800-1932* (Johns Hopkins University Press, 1984).

203 "The Model A was presented in late 1927": Ford, Model A history and River Rouge history describe the retooling interval.

PART IV

207 "Their days are imagined": Part IV's Austin, orbital-yard and Mars scenes are illustrative possibilities built from working systems, active trials and announced designs. The pictured days have not occurred.

CHAPTER 13
EARTH, RE-STACKED

209 "It had no driver and no steering wheel": Tesla unveiled the Cybercab robotaxi at the October 2024 "We, Robot" event; on the Q4 2025 earnings call, January 28, 2026, the company described a limited paid driverless service in Austin with "well over 500" vehicles and no safety monitor — company-reported deployment language rather than a regulator-certified measure of broad commercial operation.

209 "an Optimus was crossing the street": Tesla Q4 2025 earnings call, January 28, 2026; Optimus Gen 3 was described as the first design intended for mass production, with production planned before the end of 2026, an eventual target of one million robots per year, and a stated target price of roughly $20,000 to $25,000 at volume — company targets rather than demonstrated production.

209 "a row of Megapack units": Tesla Megapack product pages and quarterly update decks; each Megapack stores roughly 3.9 megawatt-hours and delivers 1.9 megawatts, and Tesla deployed 46.7 gigawatt-hours of storage in 2025.

210 "a small Starlink terminal angled at the sky": SpaceX Form S-1, May 2026, as amended; Starlink reached 10.3 million subscribers as of March 31, 2026, with service in 164 countries, territories, and other markets and median latency around 25 milliseconds.

212 "watched the driverless Semis run those same roads": Tesla's Semi program provides the working base for this future scene; driverless service remains an extension of the program, not an observed deployment.

213 "The warehouse ran on Megapack-buffered solar, charging its": American Clean Power / Tesla (https://www.axios.com/2024/10/11/tesla-robotaxi-tesla-cybercab-we-robot-musk), Mike Snyder energy-storage leadership and projected factory setting; American Clean Power / Tesla (https://www.tesla.com/megapack), Mike Snyder energy-storage leadership and projected factory setting; American Clean Power / Tesla (https://cleanpower.org/about/team/michael-snyder/), Mike Snyder energy-storage leadership and projected factory setting

214 "the family of chips that drove the car": SpaceX Form S-1, May 2026, as amended; the Terafab initiative with Tesla, joined by Intel in April 2026, targets one terawatt per year of compute hardware, including a chip type optimized for terrestrial edge and inference "to be used primarily in Tesla's Optimus robots and vehicles."

214 "moved through ground-station-to-orbit-to-ground links": SpaceX Form S-1, May 2026, as amended, and official Starlink technical materials; roughly 9,600 broadband and mobile satellites in low Earth orbit as of March 31, 2026, with optical inter-satellite links letting the constellation operate as a routed mesh.

216 “train and retrain the models that ran everything else”: SpaceX Form S-1, May 2026, as amended; the Colossus build record — approximately 100,000 H100 processors brought online in 122 days, with later clusters in 91 and 64 days and roughly one gigawatt of combined compute — is the disclosed basis for the training layer beneath the scene.

217 “There had been a time when a factory”: The Henry Ford, Rouge Factory History and Timeline (https://www.thehenryford.org/collections/explore/popular-research-topics/ford-rouge-factory-tour-history-and-timeline), Rouge complex employed more than 100,000 at peak across many buildings

218 “The materials moved in loops as much as”: International Aluminium Institute / Aluminum Association (https://international-aluminium.org/landing/aluminium-recycling-saves-95-of-the-energy-needed-for-primary-aluminium-production/), Approximately 95 percent energy savings from aluminium remelting; International Aluminium Institute / Aluminum Association (https://www.aluminum.org/Recycling), Approximately 95 percent energy savings from aluminium remelting

218 “pulled out at the end of each life”: JB Straubel, *Ride the Lightning* podcast, Episode 501, March 2025; Straubel founded Redwood Materials in 2017 and describes it as “a second almost unmanufacturing plant” built toward a closed-loop ecosystem in which battery materials are recovered and returned to production.

220 “Megapack farms inhaled the midday excess”: Tesla quarterly update decks and investor materials; storage deployments grew from 3.0 gigawatt-hours in 2020 to 46.7 gigawatt-hours in 2025, with the Lathrop and Shanghai Megafactories each at a company-stated 40 gigawatt-hours of annual capacity.

223 “There had been mornings, not many years before”: FERC (2021) (https://www.ferc.gov/news-events/news/

final-report-february-2021-freeze-underscores-winterization-recommendations), 2021 Texas freeze: failed generation, fuel supply and winterization

223 "the numbers had been falling for years": BloombergNEF battery price survey and IRENA renewable cost data; battery pack prices fell from approximately $1,400 per kilowatt-hour in 2010 to roughly $108 by 2024, and utility-scale solar's levelised cost fell from $0.417 per kilowatt-hour in 2010 to $0.043 in 2024.

CHAPTER 14
ORBIT AS INDUSTRY

225 "carrying optical links that let it talk to its neighbors": Official Starlink technical materials describe optical inter-satellite laser links that let the constellation operate as a routed mesh, with satellites in a low shell at roughly 550 kilometers.

226 "The cadence had become weather": The Vandenberg morning is a projected scene. SpaceX Form S-1, May 2026, records the fleet's repeat-launch pattern, including 2025 mission and booster-reuse figures; the particular future morning is constructed.

226 "Gwynne Shotwell had been running the operating side": Limitless Space Institute, Gwynne Shotwell profile records her presidency and day-to-day operating role since 2008.

226 "flown and reflown until the flying was the cheap part": SpaceX Form S-1, May 2026, as amended; Falcon 9 has demonstrated 34 reflights of a single first-stage booster, and of 170 missions in 2025, 159 flew on flight-proven boosters with better than 99 percent recovery success.

226 "the large majority of everything humanity sent toward orbit": SpaceX Form S-1, May 2026; approximately 2,200 metric tons launched in 2025, over 80 percent of the world's mass to orbit.

227 "A shared manifest let those builders plan around a known ride": NASA, *Small Spacecraft Technology State of the Art*, §10.3.2 describes scheduled Falcon 9 Transporter rideshare opportunities.

227 "Their radio signals also bounced off the sea": NASA, CYGNSS mission and NASA/JPL, GPS-reflection account explain how reflected navigation signals reveal ocean-surface roughness and wind. The proposed launch customer is illustrative.

228 "The satellites themselves came off a line. The": SpaceX EU Prospectus (5 June 2026), Redmond manufacturing section (https://content.spacex.com/cms-assets/FINAL_Documents%20and%20Updates/SpaceX%20-%20EU%20Prospectus%20%28Approved%20by%20Bafin%29%20-%20June%205%2C%202026.pdf), Redmond factory averaged approximately 70 satellites per week from December 2025 through April 2026

229 "tens of thousands of satellites in low orbit": FCC approval in January 2026 added 7,500 Gen2 satellites, lifting Starlink authorization to roughly 15,000, with a full configuration target of 42,000 based on SpaceX regulatory filings.

229 "built to work for about five years": FCC, Gen2 Starlink authorization, FCC-22-91A1 describes a five-year satellite design life.

230 "the dense daily load still ran through fiber": FCC 26-55 (https://docs.fcc.gov/public/attachments/FCC-26-55A1.pdf), satellite broadband performance/capacity discussion, finds satellite service behind major fiber and cable operators and identifies capacity constraints in high-demand areas. SpaceX Direct-to-Cell update (https://www.spacex.com/updates/reusability) describes mobile coverage as augmenting high-capacity terrestrial 5G networks. Starlink Transit (https://starlink.com/business/transit) says local providers distribute satellite capacity over last-mile fiber, fixed wireless or mobile networks.

230 "a standard handset with no special antenna": SpaceX Form S-1, May 2026; Starlink Mobile served about 7.4 million monthly unique devices across roughly 30 countries through approximately 650 V1 Mobile satellites; Starlink, T-Mobile, and FCC materials describe direct-to-cell as supplemental coverage for standard phones.

230 "Airlines fitted the terminals and sold connectivity at altitude": SpaceX Form S-1, May 2026; United Airlines, Carnival, and Maersk are named among Starlink customers, alongside John Deere.

231 "Tens of millions of subscribers paid a monthly bill": SpaceX, Form S-1 (May 2026), reported 10.3 million Starlink subscribers as of 31 March 2026. "Tens of millions" describes the chapter's later projected scene.

231 "the whole loop turned on its own without a subsidy": SpaceX, Form S-1 (May 2026), reports that Connectivity segment Adjusted EBITDA was positive from 2023, with $11.4 billion of 2025 revenue and $7.2 billion of segment Adjusted EBITDA. The self-funding renewal loop is a later scenario, not an established cash-flow result.

231 "Filling low orbit with tens of thousands of": SpaceX 2026 S-1 (https://www.sec.gov/Archives/edgar/data/1181412/000162828026036936/spaceexplorationtechnologi.htm), More than 1,000 automated collision-avoidance maneuvers per day in 2025; Starlink Collision Avoidance CONOPS (https://space-safety.starlink.com/docs/space-safety-articles/starlink_cola_conops/), Autonomous maneuvers with human triage for off-nominal cases

231 "Astronomers complained that the bright trails crossed": IAU Centre for Protection of Dark and Quiet Sky position (https://cps.iau.org/documents/42/Draft_IAU_CPS_position_statement_9Z8xEO1.pdf), Astronomical interference and mitigation discussion; Starlink brightness mitigation practices (https://www.starlink.com/public-files/Brightn

essMitigationBestPracticesSatelliteOperators.pdf), Operator-side brightness mitigation

232 "a connection fast enough to bring its services": Starlink Specifications (https://www.starlink.com/legal/documents/DOC-1431-92252-65) list typical broadband speeds and caution that performance varies with usage and conditions. FCC 26-55 compares satellite performance with fiber/cable. The remote village is a projected illustrative scene, not a measured equal-speed comparison.

232 "A Starlink bus could carry a different kind of payload": Reuters, report on SpaceX AI satellites (9 June 2026) describes the Starlink V3 lineage and removal of broadband antennas. SpaceX's Starmind page describes the planned AI1 system.

232 "up to 250 kilowatts of peak compute power and 175 kilowatts": SpaceX, Starmind design table lists up to 250 kW peak and 175 kW average compute power per proposed AI1 satellite. The cited Starmind description was checked on 6 October 2026; it is distinct from the June interview's earlier aims.

233 "machines were already making Starlink user terminals": SpaceX, "Watch @ElonMusk provide a technical update on SpaceX's capability to manufacture, launch, and operate AI satellites at scale," X video post, 8 June 2026 (https://x.com/SpaceX/status/2064099405758906727), 08:44–09:52: Bastrop terminal production next to proposed satellite building; no precise parcel boundary or October construction status established.

233 "A solar manufacturing facility was under construction first": SpaceX, "Watch @ElonMusk provide a technical update on SpaceX's capability to manufacture, launch, and operate AI satellites at scale," X video post, 8 June 2026 (https://x.com/SpaceX/status/2064099405758906727), 09:09–09:26: solar manufacturing described as under construction in June; AI-satellite building to follow.

233 "solar technology intended for Starlink V3": SpaceX, "Watch @ElonMusk provide a technical update on SpaceX's capability to manufacture, launch, and operate AI satellites at scale," X video post, 8 June 2026 (https://x.com/SpaceX/status/2064099405758906727), 04:54–05:07: solar technology intended for Starlink V3 described as basis for larger satellite system.

233 "as soon as late 2027 and build thousands": SpaceX, Starmind Gigasat section describes a plan for thousands of AI satellites beginning as early as late 2027. The cited Starmind description was checked on 6 October 2026; it is distinct from the June interview's earlier aims.

233 "an annualized gigawatt a year of orbital compute": SpaceX, "Watch @ElonMusk provide a technical update on SpaceX's capability to manufacture, launch, and operate AI satellites at scale," X video post, 8 June 2026 (https://x.com/SpaceX/status/2064099405758906727), 13:01–13:58: end-2027 one annualized gigawatt per year framed as an aim, with later scale aspirations. SpaceX EU Prospectus, 5 June 2026, §5.5.8.3 p.128 (https://content.spacex.com/cms-assets/FINAL_Documents%20and%20Updates/SpaceX%20-%20EU%20Prospectus%20%28Approved%20by%20Bafin%29%20-%20June%205%2C%202026.pdf), gives a long-term 100 GW annual launch goal without a deadline in that passage.

233 "as many as a million AI satellites": SpaceX, "Watch @ElonMusk provide a technical update on SpaceX's capability to manufacture, launch, and operate AI satellites at scale," X video post, 8 June 2026 (https://x.com/SpaceX/status/2064099405758906727), 07:48–07:52: up to one million satellites discussed as possible scale, not approved fleet.

235 "transferred several metric tons of cryogenic propellant": NASA, Artemis report on Starship's IFT-3 describes an internal cryogenic transfer demonstration. This was not yet ship-to-ship refueling.

235 "Kathy Lueders, who ran the Starbase operation": 2024 Starbase incorporation letter reported by AP (https://apnews.com/article/d251a58f32797fd04e4215a4570e782f), Lueders signed as Starbase general manager in 2024; later continuation is constructed

236 "to land its astronauts on the lunar surface": NASA Human Landing System awards to SpaceX, per official NASA award announcements; the Option A contract at $2.89 billion and Option B at approximately $1.15 billion for a lunar-lander variant of Starship.

237 "once about every twenty-six months": NASA/JPL, Mars launch-window lesson gives the approximate 26-month launch cadence.

237 "Planetary Resources and Deep Space Industries promised mines": Space.com, Planetary Resources acquisition and GeekWire, Deep Space Industries acquisition document the first wave.

238 "including AstroForge and Karman+": Y Combinator, AstroForge profile and Karman+, company report document the later firms.

238 "The United States and Luxembourg had drawn rules": U.S. Commercial Space Launch Competitiveness Act (2015) and Luxembourg space-resources law (2017) address rights in extracted space resources.

239 "Two hundred and fifty-nine days at minimum-energy transit": NASA/JPL, Mars launch-window lesson describes an approximately 259-day minimum-energy transfer.

CHAPTER 15
MARS AND THE KARDASHEV SCALE

241 "Late on a weeknight in 2027": The Hawthorne simulation and its exact repetitions are an illustrative future scene. SpaceX public Mars timeline material, including 2016 and 2017 presen-

tations, supplied earlier schedule context; public statements in early 2026 placed cargo flights closer to 2030. The scene is not evidence that the described 2027 room or run occurred.

242 "Descent remained one of the hardest parts of": NASA Science Mars landing explainer (https://science.nasa.gov/solar-system/skywatching/night-sky-network/landing-on-mars-a-tricky-feat/), Parachutes successful for robotic landers but inadequate alone for large landing mass

244 "a dozen light-minutes away or more": NASA and Jet Propulsion Laboratory materials place the one-way Earth-Mars signal delay at roughly 3 to 22.4 minutes depending on orbital position, and describe Mars missions' reliance on relay orbiters and the Deep Space Network.

244 "The early manifests carried machines before people": SpaceX public statements and Musk presentations, including the 2016 International Astronautical Congress address, describe cargo flights preceding crews; SpaceX statements in February 2026 placed Mars cargo flights closer to 2030.

245 "Bill Gerstenmaier had spent much of his career": U.S. Senate Commerce Committee / AIAA (https://www.commerce.senate.gov/wp-content/uploads/media/doc/William%20Gerstenmaier%20Written%20Testimony%20Before%20Senate%20Commerce%2010.16.2023.pdf), Bill Gerstenmaier NASA human-spaceflight role and later SpaceX role; U.S. Senate Commerce Committee / AIAA (https://aiaa.org/people/william-gerstenmaier/), Bill Gerstenmaier NASA human-spaceflight role and later SpaceX role

246 "The chemistry was old. A French chemist named": Nobel Prize / IUPAC (https://www.nobelprize.org/prizes/chemistry/1912/sabatier/lecture/), Paul Sabatier's catalytic chemistry and 1912 Nobel award; Nobel Prize / IUPAC (https://publications.iupac.org/ci/2013/3505/2_lattes.html), Paul Sabatier's catalytic chemistry and 1912 Nobel award

247 "The dust was the reason the reactors kept": NASA/JPL, "Mars Climate Sounder Studies 2018 Dust Storm," 30 October 2018 (https://www.jpl.nasa.gov/images/pia22737-mars-climate-sounder-studies-2018-dust-storm/), description paragraphs 1–3: 2018 global storm affected most of Mars, stayed near peak three weeks, then decayed over months; tau 3 lets about 5% of direct sunlight reach surface. NASA/JPL, "Six Things About Opportunity's Recovery Efforts," 2018 (https://www.jpl.nasa.gov/news/six-things-about-opportunitys-recovery-efforts/): planet-encircling dust cut solar power to Opportunity. Storm timing in the passage is general, not a claim of periodic regularity; the settlement power implication is scenario reasoning.

249 "The scale of it was hard to hold": SpaceX Starship specifications (https://new.spacex.com/vehicles/starship), Current Starship page gives 1,600-tonne vehicle propellant capacity. The roughly 1,000-tonne Mars return-manifest target in prose is a constructed design assumption, not a validated Mars-ascent requirement.

249 "The choice of methane went back to the": FlightGlobal (https://www.flightglobal.com/spacexs-mars-rocket-to-be-methane-fuelled/107953.article), Methane/oxygen architecture for Mars propellant production

251 "In the latter part of the twentieth century": NASA / Purdue University (https://www.nasa.gov/mission/apollo-11/), Buzz Aldrin's Apollo 11 role and cycler advocacy; NASA / Purdue University (https://engineering.purdue.edu/AAC/exploration-projects/), Buzz Aldrin's Apollo 11 role and cycler advocacy

251 "What made the idea worth reviving was what": NASA human body in space (https://www.nasa.gov/humans-in-space/the-human-body-in-space/), Radiation and bone/muscle loss during long-duration flight

252 "Bret Drake and other Mars-architecture planners had studied": NASA Mars Design Reference Architecture (https://www.nasa.gov/pdf/373665main_NASA-SP-2009-566.pdf), NASA Mars architecture and cycler-orbit discussion

254 "The solar system set the timetable. Earth and": NASA JPL / SpaceX (https://www.jpl.nasa.gov/edu/resources/lesson-plan/lets-go-to-mars-calculating-launch-windows/), Earth-Mars launch-window cadence of about twenty-six months; NASA JPL / SpaceX (https://www.spacex.com/humanspaceflight/mars), Earth-Mars launch-window cadence of about twenty-six months

254 "There was an old scale for thinking about": Kardashev 1964 paper (https://ui.adsabs.harvard.edu/abs/1964SvA.....8..217K/abstract), Original first-kind reference ~4 x 10^12 W; later 10^16 W convention is distinct

254 "In 1964 a Soviet astronomer named Nikolai Kardashev": Nikolai S. Kardashev, "Transmission of Information by Extraterrestrial Civilizations," *Soviet Astronomy*, Vol. 8, No. 2 (1964); originally published in Russian in Astronomicheskii Zhurnal, March-April 1964, and presented at the Byurakan Conference on Extraterrestrial Civilizations in Soviet Armenia.

254 "His first kind could command power on the order": Nikolai Kardashev, "Transmission of Information by Extraterrestrial Civilizations," Soviet Astronomy 8 (1964), pp.217–221 (https://adsabs.harvard.edu/pdf/1964SvA.....8..217K), p.217: original Type I is comparable to then-current Earth power, approximately 4 × 10^12 W. Carl Sagan, The Cosmic Connection (1973) (https://archive.org/details/cosmicconnectionoosaga), later gives a continuous reading. Do not conflate the original Type I with the later 10^16 W convention.

255 "The human world still drew much of its power": Energy Institute, Statistical Review of World Energy 2025, media release 26 June 2025 (https://www.energyinst.org/exploring-energy/resources/news-centre/media-releases/

renewables-soar%2C-but-fossil-fuels-continue-to-rise-as-global-electricity-demand-hits-record-levels), Key Findings: 2024 world energy demand reached 592 EJ; all fuel categories rose, including fossil fuels. The prose is a qualitative energy-mix statement, not a numerical fraction of intercepted sunlight.

255 "Carl Sagan, who did as much as anyone": Sagan, The Cosmic Connection (1973) (https://archive.org/details/cosmicconnectionoosaga), Introduced logarithmic interpolation; historical estimate around 0.7 is distinct from present input; Energy Institute Statistical Review 2025 (https://www.energyinst.org/exploring-energy/resources/news-centre/media-releases/renewables-soar%2C-but-fossil-fuels-continue-to-rise-as-global-electricity-demand-hits-record-levels), 2024 global energy demand 592 EJ/year, ~18.8 TW; Sagan formula (log10(P in W)-6)/10 gives ~0.73, not 73 percent of Type-I energy

255 "smoothed the three steps into a continuous scale": Carl Sagan, *The Cosmic Connection: An Extraterrestrial Perspective* (Anchor Press/Doubleday, 1973), which popularised a decimalised version of the scale and placed contemporary humanity below the first type.

256 "Musk imagined making solar arrays and radiators": SpaceX Form S-1, May 2026, as amended; the filing describes lunar factories using lunar resources to manufacture millions of AI compute satellites, with electromagnetic launch from the lunar surface described under planned capabilities.

257 "The hedge had a fossil record behind it": NSF / Natural History Museum (https://www.nsf.gov/science-matters/moment-changed-earth), Chicxulub impact and extinction evidence; NSF / Natural History Museum (https://www.nhm.ac.uk/discover/how-an-asteroid-caused-extinction-of-dinosaurs.html), Chicxulub impact and extinction evidence

CHAPTER 16
THE STACK OUTLIVES THE FOUNDER

260 "Other layers are at different stages. The robots": Tesla 2025 Form 10-K (https://www.sec.gov/Archives/edgar/data/1318605/000162828026003952/tsla-20251231.htm), Optimus described as a humanoid robot in development

261 "was founded in 2002": SpaceX Form S-1, May 2026; the filing's opening business description begins "Founded in 2002, SpaceX is the only company building the integrated hardware and software infrastructure of the future across space, connectivity, and AI."

261 "the first Falcon 1 reached orbit only after three failed attempts": NASA, *Aeronautics and Astronautics: A Chronology, 2008*, p.42 records Falcon 1's first successful orbital flight after three failed attempts.

262 "later turned to recovering the metals from cells": JB Straubel, interviewed by Ryan McCaffrey, *Ride the Lightning* podcast, Episode 501, March 2025; Straubel left Tesla in 2019 and founded Redwood Materials, a battery recycling and remanufacturing company he describes as "a second almost unmanufacturing plant."

262 "Engineers who worked on the cars' vision": DensityAI founder statement (https://www.linkedin.com/posts/densityai_building-a-product-is-fun-building-one-that-activity-7473090853201354754-fTmm), Ganesh Venkataramanan identifies Tesla Dojo work

262 "Gwynne Shotwell had spent the better part": Northwestern University / TIME (https://findingaids.library.northwestern.edu/agents/people/2758), Gwynne Shotwell's SpaceX presidency, operating role and tenure; Northwestern University / TIME (https://time.com/article/2026/03/26/spacex-gwynne-shotwell-full-interview/), Gwynne Shotwell's SpaceX presidency, operating role and tenure

ABOUT THE AUTHOR

Reid J. Osterley writes about industrial systems and the people who build them. His subjects range from launch vehicles and power grids to factories and the supply chains that connect them. He works from the public record, drawing on securities filings, regulatory dockets, court documents, and the published statements of the companies at the center of his work. *The Civilization Stack* is his first book.

ENJOYED THE BOOK?

Thank you for reading. If this book gave you a clearer view of the machine being assembled in plain sight, would you take a minute to leave a review on Amazon? Honest reviews help other readers find the book, and they make a real difference for a small, independent publisher.

SCAN TO LEAVE A REVIEW

amazon.com/dp/1968988653

Printed in Dunstable, United Kingdom

91219626R00193